王劲韬 著
By Wang Jintao

Basic Training

基础训练

景观设计手绘教程
Hand Drawing
for Landscape Design

中国建筑工业出版社
CHINA ARCHITECTURE & BUILDING PRESS

前　言

　　本丛书共三册，其框架结构是本着一种解剖麻雀和组装机器的思路，也就是笛卡儿所说的分解与结合，将效果图中各元素的提炼、提升和相应的技法运用，视为组装一幅成功效果图的必要零件。只有把每一个零件或模块画得既快又好，做到信手拈来，才有可能"组装"出一幅成功、感人的效果图作品。

　　丛书第一册从最简单的一个树圈开始，到树的不同种类形态、光影、季相的表达，再到天空、水面、建筑的综合画法，对景观表现图中常用题材，特别是各种树木的具体技法做了详细展示，通过大量实例演示，为初学者提供较多的参照学习样板。以例说法，学习效果图"零件"的表现，通过每一棵树、每一块石、每一片水面的临摹练习，达到谙熟于心的地步，从基础的细节层面，一砖一瓦地构建起景观设计表现的"大厦"。

　　作为初学设计表现的同学，最重要的是在快与好之间做出很好的平衡，即快而不流于草率，精而不流于拖沓。书中的手绘图无一例外地采用了彩铅、马克笔等简单易行的表现工具，高度重视了普通的彩铅（除了极精细的绘画，作者认为普通建筑景观设计中，无需水溶性铅笔）与马克笔混合使用的综合技法，这些材料是快速制图的最佳选择。景观设计表现从本质上讲是一种快中求好、快中求准的艺术，唯有快速概括场景和设计的本质才能抓住瞬息万变的设计机遇和转瞬即逝的设计灵感，其余诸如精雕细刻的场景渲染和一丝不苟的再现工作，在当代的设计过程中，完全可由电脑去完成。

Foreword

This series comprises of three volumes, with a framework based on the logic of "dissecting a sparrow" and "assembling a machine", that is, what Descartes called decomposition and combination. Refining and improvement of elements and application of skills can be considered as the "parts necessary to assemble" a successful rendering. A successful and appealing rendering work won't be "assembled" unless the drawer is able to present each part or module both rapidly and in good quality, with everything at hand for picking up.

Starting from a simple "tree circle", Volume I presents in detail the themes commonly used in landscape presentation drawings, and especially specific techniques to present trees, covering different kinds and forms of trees, the expression of light and shadow, as well as seasonal appearances, and the combined techniques to paint the sky, water surface and buildings. Plenty examples are given as the reference models for beginners, through which they can learn how to present the "parts" of a rendering. By copying every tree, rock and water surface, they will have a firm grasp on the techniques, and finally build their own structure for landscape design presentation from every small basic detail.

For design presentation beginners, the most important thing is to balance between speed and quality, so as to be fast rather than hasty, and refine rather than procrastinate. In all hand drawings in the book, simple tools such as color pencils and markers are used. The combination of common color pencils and markers is highly valued (except for very delicate painting, because the author believes that water soluble color pencils are not necessary in ordinary architectural landscape design). These are the best choices for presenting drawings both rapidly and well. Landscape design presentation is essentially an art that seeks quality and precision from speed, where the opportunities are ever-changing, and inspirations for design are sparkling at a glimpse, both fading away unless you are able to capture the essence of the scenes and design. As for other tasks such as intricate scenario rendering and meticulous on-line works, just leave them to the computer, which is a helpful assistant in modern design to get these works well done.

目 录

010	**1**	**导言：当代景观规划语境下的手绘图**
010	1.1	作为说明和交流工具的手绘图
012	1.2	作为设计思维过程的展现和具体化的手绘图
016	1.3	手绘图的未来
020	**2**	**景观元素的表达**
020	2.1	树的表达
053	2.2	天空的表达
060	2.3	水的表达
068	2.4	山石的表达
076	2.5	建筑物的表达
084	2.6	人物的表达
094	2.7	车和船的表达
098	**3**	**景观色彩的表达**
098	3.1	植物色彩的表达
120	3.2	天和水色彩的表达

Contents

010   Chapter I Introduction: Hand Drawing in Modern Landscape Planning

010   1.1   Hand Drawing as A Tool of Illustration and Communication

012   1.2   Hand Drawing as A Specified Presentation of Designing Process

016   1.3   The Future of Hand Drawing

020   Chapter II Expression of Landscape Elements

020   2.1   Expression of Trees

053   2.2   Expression of the Sky

060   2.3   Expression of Water

068   2.4   Expression of Hills and Rocks

076   2.5   Expressions of Architecture

084   2.6   Expression of People

094   2.7   Expression of Vehicles and Vessels

098   Chapter III Expression of Landscape Colors

098   3.1   Expression of Plant Colors

120   3.2   Expression of Sky and Water Colors

1 导言：当代景观规划语境下的手绘图
Chapter I Introduction: Hand Drawing in Modern Landscape Planning

1.1 作为说明和交流工具的手绘图

1.1 Hand Drawing as A Tool of Illustration and Communication

手绘图是设计师自己的交流工具，也是设计师与业主的交流工具。建筑画作为建筑设计发展史上的一个阶段性的成果，用以评价或者阐述一些观点，这一手法恐怕是从文艺复兴就开始了。比如达·芬奇手稿中列举的那些穹顶、梁托，著名的维特鲁威人图示等，达·芬奇手稿中的那些草图似乎更多的是作者与自己的思想做交流的过程展示，通过草图描述并记录自己的思考轨迹，所以毋宁称之为笔记。而伯拉孟特为圣彼得教堂所做的那些精准的平面和立面草图则更多地带有向甲方业主（教皇克莱门七世）阐述其空间理想的意味。

在景观建筑表现图的历史上还有一类纯粹理想化的图纸，这是独立于以上两类的特例。诸如英雄主义建筑师艾蒂安·布雷的大都市教堂和牛顿祠，以及浪漫主义大师辛克尔早年的一些舞台设计，如他为沃夫冈·莫扎特的歌剧《魔笛》所做的极富象征意义和历史主义隐喻的舞台场景。这种浪漫主义的艺术追求到了20世纪，在渲染大师沙勒尔的许多方案表现图中仍然得到了充分体现。直至近代美国人赖特的作品——伊利诺伊摩天大厦的渲染图，似乎都是在阐述一种理想和对未来的一种预言。这类作品更类似于今日的动漫设计，诸如宫崎骏的动画，抑或技术理性下的变形金刚等形象。电脑软件的场景渲染技术在这一方向上所展示的超能力，已经使这类作品完全不需要手工的干预（除了最初的基本形），其作品感染力也大大超过了当年辛克尔创作《魔笛》这一类场景所能达到的水平。

Hand drawing, not only serves as a tool of communication for a landscape designer herself, but also a tool of communication with clients. Architectural drawing is a periodic achievement in the history of architecture design, which is used to comment or depict ideas. Perhaps this technique can be traced back to the Renaissance, such as the domes and bolsters in the drawing of Leo da Vinci, and the famous Uomo Vitruviano. The drawings of Leo da Vinci are more like an illustration of the process that the drawer communicated with himself. Drawing was used to describe and record the tracks of his thoughts, thus rather a note than a drawing for him. As to the precisely drawn plan and elevation sketches of the Basilica di San Pietro in Vaticano by Donato Bramante, they are more inclined to depict space ideal for the client (Pope Clement VII).

In the history of landscape architecture presentation drawing, there is another kind of drawing — the purely idealized drawing which distincts from those two kinds. Such kind includes Metropolitan Cathedral and Newton Memorial of Heroism architect Etienne Louis Boullee, and some stage designs of Romanticist Master Karl Friedrich Schinkel in his early years, e.g. the stage scenes he designed for Wolfgang Amadeus Mozart's opera "The Magic Flute", which were full of symbolic and historical connotations. Even in the 20th century, the artistic pursuit of Romanticism was still found overwhelming in many plan presentation drawings of rendering master Schaller, let alone American architect Frank Lloyd Wright's The Mile High Illinois, whose rendering seems no more than depicting of an ideal and farseeing the future. These works are more similar to cartoon design, such as Miyazaki Hayao's animation, or the figures like Transformers in the context of rational techniques. In fact, the rendering techniques of computer software performed a presentation so great that this kind of works can be kept off any hand drawing (expect the basic form), and the final works are even much more appealing than such scene as Schinkel designed for "The Magic Flute".

伊利诺伊摩天大楼：1956年计划在美国芝加哥建造。这座摩天大楼的高度设计为1609米，设计师是弗兰克·劳埃德·赖特，他认为当时提议建造这样的宏伟建筑是可能实现的。这座摩天大楼包括528层，总面积达到1714990平方米。伊利诺伊摩天大楼方案提出后发现一系列问题，比如：维护电梯的空间将占据较低楼层的所有可用空间等。因此，这项建造摩天大楼的计划被迫夭折。

所以我说，今日手绘，撇开与电脑在技法优劣方面的对比，其发展方向已从过去追求尽善尽美的光影表达和环境渲染，转向另一种境界的追求——对创作意向尽可能迅速地把握、抓取和记录，进而推向一种表达瞬间意念的方向。事实上，许多稍纵即逝的设计构思唯有通过快速手绘图，才有可能保存和记录下来，这种捕捉瞬间意念的能力是在电脑上不可能实现的。清华大学的朱文一教授曾就第一届建筑手绘图大赛撰文提出一种特殊的手绘工具——手鼠笔，这是一种将鼠标和传统画笔完美结合的超一流的快速设计工具，可以在电脑屏幕前高速模拟钢笔、铅笔、水彩、色粉甚至油画等各种机理和特效。但实质问题是，这一切特效还是来自于作者的手绘创作，还是手脑配合的产物，只是高科技可以使我们站在更高的平台上，用比传统绘画多上几千倍的色系选择进行创作。但究其根本，仍然没有否定手工绘图在设计、尤其是创作阶段的重要价值。朱文一先生用幽默的语言表达了目前作为中流砥柱的一代中青年建筑师对于传统手绘继承和现代电脑渲染技术应用这两者关系的一种主流观点：将电脑技术与传统手绘相结合，更好地抓住设计师的灵光一现。

此外，在艺术化地表现创作思维领域，手绘图同样具有不可替代的优势，这也是手绘图作为交流工

The Mile High Illinois, was planned to be built in Chicago in 1956. It was designed by Frank Lloyd Wright, who thought such huge building could be built at that time. With a total height of 1,609m, the skyscraper had 528 floors, and the total area amounted to 1,714,990m^2. However, a series of problems rose afterwards, for instance, all the available space of the lower floors had to be used for elevator maintenance. Therefore, the project had to be terminated.

So I think that hand-drawing (despite simple technical disadvantages vs. advantages), instead of pursuing perfect light-shadow expression and scenery rendering, are heading to the pursuit of another stage nowadays: to seize, grasp and record the inspiration as quickly as possible, and to present the flowing ideas in a twinkle. In fact, only the quick sketch can keep a record of many instant ideas, which cannot be realized by computer. In his essay on the 1st Architectural Hand Drawing Contest, Mr. Zhu Wenyi, professor of Tsinghua University, put forward his ideal quick design tool which combines mouse and traditional paintbrush perfectly and can simulate the mechanism and special effects of pen, pencil, watercolor, gouache, and even painting. It is only that high-tech raises us up to a higher platform where creating with thousands more colors than traditional drawing is available. However, all those special effects still originate from the illustrator's hand drawing creation and come from his hand-brain cooperation, and the essential part of this imaginary tool never denied the important role that hand drawing played in design, and especially in the creation process. Mr. Zhu Wenyi's humorous language expressed an mainstream view of the present relationship between the inheritance of traditional drawing and the application of modern computer rendering techniques: to combine the computer techniques with traditional drawing, so as to catch the instant inspiration in design.

Besides, in the field of artistically presenting creation thinking — also another important feature of hand drawing as a communication tool — hand drawing shows its irreplaceable advantage. The style, form, and personal temperament presented in a drawing is more

具的一个重要特征。我们在手绘表现图里所呈现的笔法形式、主观情趣等内容往往更多的是在传达设计者在设计品位、设计兴趣和关注点等方面的信息，而不是亦步亦趋的实景模拟，这往往是解读这些手绘"天书"的前提。这就有点像达·芬奇笔记中的那些看似潦草的说明图，在承认作者不凡的艺术灵气和激情的同时，更需要解读作者用隐晦的方式传达出的设计信息和趣味焦点。在纯建筑领域，阿道罗西是这方面的典范，他的建筑草图在我看来是用孩子一般的天真烂漫表达出严谨缜密的建筑思维。作为一种独立的艺术表达形式，大师的手绘草图往往寥寥数笔就能表达出意味无穷的境界，因为这寥寥数笔包含了作者深厚的学识素养和对设计项目深刻的理解，这就是所谓的功夫在画外。这应该是建筑手绘图在自我交流和团队交流方面的第二个发展方向。

1.2 作为设计思维过程的展现和具体化的手绘图

设计过程是一个思考的过程，也是一个寻找最合适表达途径的过程。在这一过程中，电脑和手绘所承担的角色应该是有所区分的。电脑的优势在于表达最终成果，其光、色及环境氛围的逼真程度是手工渲染望尘莫及的。但是这仅止于最后阶段，电脑渲染技术无疑为景观建筑表现艺术开辟了更加广阔的前景，但当前的问题在于，电脑渲图技术在大获发展的同时，带来了设计过程中的各种"变异"，简言之即以"表现"代替"设计"。这种趋势发展下去无疑是相当可怕的：用软件研究代替空间研究，用数字堆砌代替艺术品质的推敲，说明文本完全为各式"美图"所替代，进而设计成果展示几乎异化为竞图比赛等等。于

a delivery of the information about the illustrator's taste, interest, and concerning points in design than the rigid simulation of realistic view. And this often serves as a premise to interpret these "wordless" works. Just like the seemingly hasty illustrations in Da Vinci's notes. Besides acknowledging his outstanding art ingenuity and passion, it counts more to know the referring information and focus of interest which are delivered in an obscure way. In the field of pure architecture, Aldo Rossi is a role model. His architectural sketch, from my point of view, though seemed as simple and innocent as a child, contained strict and meticulous architectural thoughts. Hand drawing as an independent artistic expression, can often express endless meaning in several simple stokes by great draftsmen, for these strokes contained their rich academic possession and deep insights on the designing project. This is what we call "art of drawing coming from the outside". This should be the second development direction of architectural hand drawing with respect to inside and outside communication.

1.2 Hand Drawing as A Specified Presentation of Designing Process

Designing is a process of thinking as well as a process of finding the best way of expression. However, the role computer plays in the process should be distinctive from hand drawing. Computer has its unique advantage in expressing the final result; the verisimilitude of light, color and environment are far more perfect than hand rendering, only in the final stage, though. No doubt, computer rendering techniques open a better future for landscape expression artistry. However, with its dramatic development, it brings multifarious "variations"— namely, expression takes the place of design. It would be much more terrible if keeps growing like this: software development takes place of space research; digital generation takes place of artistic consideration; illustrative text is totally replaced by pretty pictures; finally, design presentation becomes a variant of picture competition. As to the clients, they are often attracted by pretty

甲方而言，在专业视野有限的前提下，往往无法在创作初期就触及项目的实质内容，而直接被"美图"所俘获，这种"电脑美图大全"给项目带来的不仅是设计变异、变味，而且这种变了味的所谓"设计"往往根本无法落地，成了空有一身好皮囊的花瓶和空中楼阁，这对项目的贻害是显而易见的。而对设计主体——设计师，或是那些即将成为设计师的学生而言，这种影响就更大。对景观设计过程的推敲远不如表达图经得起考量，"构思"一词更像是说"构图"，我看到越来越多的学生"设计"（包括纯研究性的竞赛设计）往往只有成图、电脑模型，而没有一张过程草图，这与设计的科学过程是极不相容的。

这种数字化"替代"、扼杀的可就远远不止那些设计美感和创意，还包括草图成型过程中的许多随机性变化的被扼杀和设计丰富性的缺失，最终导致学生在学习阶段就将发挥主观能力的权利拱手让出。随之而来的设计构思、分析、归纳能力以及最终创作能力的缺失，也就不足为怪了。这就是为什么我们的园林专业在录取分数逐年上升的情况下，学生素质、综合能力却急剧下降的原因之一。这一点，我们只需稍稍回顾一下我们的前辈大师们在学校期间的习作，两代建筑学子之间的差距就非常明了了。

艺术创作最重要的能力是想象力和感悟力。想象力的极致，正如布拉蒙特所创造的奇迹——将伟大的万神殿大穹顶"举起"，搁到圣彼得大教堂的巴西利卡之上。而感悟力，我们则需要做到像景观设计前辈布朗先生那样，时刻感悟到大地的脉搏，时时刻刻能发现土地的潜质和风景的诗意。而单纯强调电脑渲染技术、忽视手绘推敲过程、忽视手脑并用的训练，这种设计模式无疑剥夺了设计师发挥想象和创造的权力，感悟的过程更是被直接简化而近于虚无！但如果设计师被剪断想象的

pictures soon since they are unable to get the essential part of a project for their lack of specialized point of view. Unfortunately, those "pretty digital picture collection" brings no more than an abnormal variant of design which "designed" nothing but good-looking castles in the air. Its deteriorating effect is apparent. As to the mainstay of design —the designers and the designer-to-be students, it could be even worse. Their weighing in the designing process becomes far less than their consideration in way of expression. For them, "designing" is much closer to "picture composition". I have seen more and more students' "designs" (pure researching designs for competition included) are only computer models and renderings but not designing sketches. Rather than designing, they are organizing pictures, which is contradictory to the scientific process of design.

Perhaps more than sense of beauty and creation, even random changes and richness in design will be killed by the digital replacement. As a result, the students give up individual thinking when they are still learning; no surprising that they are followed by losing conceiving, analyzing, and inductive abilities and finally creativity. That's why the quality and comprehensive ability of our Chinese landscape majored students has been declining year by year, as a contrast to the rising admission line of the major. Just glimpsing back at the works our predecessors have done in their school, you'll see the gap between the two generations of architects.

Imagination and perception are the two most important elements in artistic creation. An extreme example of the former is the "miracle" Donato Bramante has created: he "raised" the vault of Pantheon and placed it onto Basilica di San Pitero. As to the latter, we can learn from landscape design predecessor Joseph E. Brown, who can always feel the impulse of the earth, find the potential of land and see the poetry of scenery. The design mode that simply emphasizes the digital rendering techniques, while ignoring the process of elaborate designing and the training of hand-mind cooperation, is no less than depriving the imagination and creation from a designer. What a designer could do if he was cut the wings of imagination? Undeniable, computer has its advantage in representation of the nature,

翅膀，他还能有什么可为之处？电脑渲染的优势在于最真实地再现自然，但也仅仅是重视视觉反馈，而想象则足以使人超越视觉的简单存在，用设计师的经验和智慧抽取场所的本质，创造超越简单存在的视觉自然，我们称之为灵光一现，而画出来的灵光一现更是相当可贵的，因为设计需要的正是这种在对普遍存在物的感悟和超越的基础上，进而再创造的过程。

我的结论很明显，即使在数字化技术高度发达的今天，主动放弃手绘图、草图、草图模型的推敲过程也是不可取的。这不仅是因为直接采用电脑进行初步设计（实际上几乎做不到）对于许多灵光一现的创意根本无法把握，更因为艺术创作的基本规律使我们相信：为人服务的空间，与文化紧密相连的表现艺术，不可能也不应该被简化为一种流水线式的"作品"！更何况在这种数字化过程中，许多细节在剥离了设计本身的模糊性、可变性特质以后，便会显得空洞而多有掣肘，许多成熟设计师是深悟其中三昧的。

具体而言，手绘图的优势表现在以下几个方面：

（1）在于方便快速地把握一些转瞬即逝的创意、符号，并通过后续的详细设计对创作意图加以完善、说明。这种方式类似于许多成熟的建筑师在餐巾纸、烟盒上所做的天书式的"草图作品"。在专业领域中，日本人安藤忠雄是这方面的榜样，安藤的许多方案是随机而来的，几乎是想到什么好点子就会及时画下来。从速写本到餐巾纸、报纸，几乎任何时候有灵感都能抓住，并以自己的理解加以表达。这种类似于天书的表达，虽然只有设计者本人能够读懂，但却足以起到记录、提示的作用，是下一步深化的重要基础。安藤忠雄画草图的意义恐怕不仅在于记录，这种绘图在情感跳跃过程中流露，据此几乎能听到建筑师的心声，生命力所倾注的图像，绝不仅是数字化生产线上

but it only focuses on the visual feedback. But imagination works go beyond visual sense. A designer with experience and wisdom can extract the essence of a place and create a visual nature that goes beyond simple existence — what we called inspiration in a twinkle. To perceive into the universal existence and to create beyond it is what design requires. It would be much cherished if the inspiration can be drawn out.

So here is my conclusion: even in the times of digital high-tech development, it would be a terrible mistake to abandon the weighing process of sketch, hand drawing and preliminary model. Not only because preliminary designing by computer (almost unrealizable actually) cannot seize the flowing Inspiration, but also the basic principle of artistic creation makes we believe that neither the space serving people nor the expression art that closely related with culture, could or should be simplified as an assembly line produced "works"; let alone the creating activity, being stripped off ambiguity and variability in the digitizing process, becoming empty and limited. And that had been deeply understood by many experienced designers.

The advantages of hand drawing are specialized in the following aspects:

(1) It is convenient for designers to grasp the twinkling creative ideas and symbols quickly, which can later be perfected and demonstrated by the following detail designs. It is particularity similar to may experienced architects' obscure "sketch works" which were drafted on napkins of cigarette packets. In specialized field, Ando Tadao is an example. Many of his schemes seem came by accident. Actually, he has a habit of drawing out any good idea that occurs to his mind any time on any materials—napkin, newspaper, etc. —and presenting the inspirations in his own way. Though this kind of presentation can only be read by the illustrator himself, it has fulfilled the role of records and reminders and laid an important foundation for the deepening step. Perhaps more than recording, Ando's sketches reflect the leaps of emotion that we can almost hear the architect's heart. Those heart-made pictures can never be produced by digital assembly lines.

才能实现的。事实上，现代设计过程比以往任何时候都少不了这种"餐巾纸式"的草图。安藤认为：给这些草图加上一些要素和具体尺寸后，就能整理成"设计画面"，后续许多工作往往属于完善和具体化过程，并不具有原创意义，决定性之举往往是这些不起眼的餐巾纸上的"灵光一现"。

（2）草图有利于促进设计思考和完形。如果我们承认设计是一个既有灵感的凸现、又有逐步完善的过程，那么草图在这一过程中则是思考的工具。从这个角度上讲，设计毋宁说是"画"出来的，设计制图与纯艺术的绘画不同点在于前者是成竹在胸，落笔在后，设计则往往是从一个快速闪现的灵感起步，通过逐步深入的思考而渐渐清晰的构思过程，草图则有利于促进这一思考和完善的过程。所以从某种意义上说，设计是"画"出来的。从大的概念性调整过程到每一个设计细节，往往都是在不停顿地画的过程中逐步完成的。不仅如此，新的灵感也会在草图完善的过程中逐步闪现出来，在这方面，个人的体会是颇为深刻的。如果缺少"画"的过程，我们的设计难免会像个干瘪的老太太。反之，业内很多艺术绘画和平面设计好手跻身于景观设计，并有所斩获，靠的也是这种手头硬功夫。

（3）作为说明性的草图更容易突显设计师的主观意念对元素、符号的概括和提炼作用，即更能抓住有目的的选择和强化某些重点部分，画面传递的信息更具有选择性、主观性，进而具有个性化艺术风格以及设计上的指向性。手绘图更容易导向一种元素更清晰、画面更简略、表达意图更明确的图式风格。尤其在工作草图阶段，这种个人风格和作品的意念化表达会使画面（设计）显示出极强的可识别性乃至可读性。许多大师的作品往往能在瞬间被辨别出来，其清晰程度令现代功能极其强大的电脑也望尘莫及。

In fact, this "napkin sketch" cannot be emphasized more any time than in modern designing process. Ando Tadao holds the view that just some elements and specified scales added, these sketches can be modified into "design presentations". Neither original nor determinative, many the following works are often the refining and specifying process. But what really matters twinkled on common napkins — the inspiration.

(2) The sketch helps the thinking and completing of the design. If design is a process involved both inspiration highlighting and gradual completing, sketch is a tool of thinking in this process. In this sense, designing is almost drafting. Different from pure artistic drawing being drawn out as an already existed image in the illustrator's mind, design usually starts with a flashing inspiration, and becomes clearer and clearer with deeper thinking. So in a way, designers, "drafting" designs from conceptual adjustment to every details designing in the process, are often made by constant drafting, which helps the thinking and completing. Moreover, usually new inspiration comes in the completing process, for which I had experienced deeply. Leaving out drafting, our designs can never be exempted from boring. An opposite aspiration story is that when many good artistic painters and graphic designers turn to landscape field, they usually succeed and lead fruitful careers, for their great proficiency in drafting.

(3) It highlights the designers' individual epitomizing and abstracting on elements, symbols, etc. The illustrative sketch, with better control of the oriented adoption and emphasize of certain parts, can deliver its message more selective and subjective, and then in a more personal style of being more directive in design. In a word hand drafting are more likely heading to a style of clearer elements, simpler frame, and more explicit presentation. Particularly in the sketching phase, this kind of ideas-contained expression of personal style works would make the picture (design) be more recognizable and more readable. Many masters' works can be recognized in a glimpse and their presentations are so clear that even the highly developed computers cannot achieve.

1.3 手绘图的未来

数字科技、电脑模拟软件使图面的精度达到了传统绘图方法望尘莫及的程度。当代设计师的工作台面上总是少不了各种各样的数字化模拟仪器,作为后期表现的主力,这种人机协作的方式和技术支持毫无疑问将为现代设计提供更广阔的空间,但这并不意味着草图纸、绘图桌将就此退出设计过程。我的观点非常明确,在设计构思的最初阶段进行全电脑生产线式的制作是不经济、不现实、不值得提倡的。为工作草图建立数字化模型不仅耗费大量的人力,而且时间成本极大。电脑图纸更适合的角色是作为一个终端产品,对于一些有价值的草图方案可以适度发展成为数字化草模和草图在设计初期并行发展,但摒弃手工草图,完全依赖数字模型必然会限制多种方案的并行发展。而且,电脑渲染图这个终端产品的质量和艺术性还是与设计者本人的艺术素养和制图水平紧密联系,而不可能完全作为一种纯智能化的产品独立存在于电脑软件中。事实上,高度发达的现代设计过程比以往任何时候更需要"餐巾纸"式的草图。

但无可否认的事实是,我们当前的建筑景观表达正逐步走向追求单一目标的、追求尽善尽美的环境渲染和绚丽的画面效果的终端式产品,而不是作为一种思考性的辅助工具。官方的设计竞标在相当程度上引导了这一发展方向,对方案的评审和最终取舍往往依据图面的表达效果,却很少关注过程性创作,许多精彩的过程性草图、工作草模被那些渲染精致、热闹非凡的电脑表现图所掩盖。这让人想起半个世纪以前,柯布西耶的那些精彩的研究性草图在官方竞图比赛中被直接废弃的悲哀经历。同样,在当代中国设

1.3 The Future of Hand Drawing

On the other hand, digital techniques and computer simulation software has raised the picture rendering so high that traditional drawing could never exceed; we can always find various digital simulation instruments on the workbench of contemporary designers. As computer serves as a major force in the final presentation no doubt that human-machine cooperation and technical support will create a wider stage for modern design, but it doesn't mean the sketching paper and drawing table will leave the design process. So my point is, it's less economic and unrealistic to adopt the computer workshop-producing in the very beginning of design, and thus unworthy to be tried (actually, to create digital model for sketches not only makes people exhausted, but also costs a lot of time). I think it would be more appropriate for computer drawing to act as a final product. Some valuable sketches could be developed into digital preliminary models, or both in a paralleled way of developing; however, if hand drawing were abandoned, the overall development would be restrained by relying on digital model totally. Anyway, computer renderings as the final product, whose quality and artistry still close link with the artistic quality and drafting capacity of the designing operator, would never exist in the software independently as a total artificial intelligent product. In fact, the highly developed modern design process requires the "napkin sketch" more than any time.

But the undeniable truth is our present architecture presentation becomes an end product which pursues a single result-environmental rendering and gorgeous visual effect as perfect as possible — rather than a thinking — aids aided tool that it should be. To a large extent the official competitive bidding of design has led the way of this abnormal trend. The bidding reviewers are often more attracted by the visual effect of presentations and less concern about the progressive creation, thus many wonderful progressive sketches and preliminary models fade before those exquisite glamorous computer renderings. It reminds us the pitiful experience of Le Corbusier half a century ago, when his

计一线，这样让人啼笑皆非的闹剧天天都在上演。由于欣赏能力和艺术、技术素养方面的限制，由于经济政治等方面的驱动，在我们这个领域，草图这一具有决定意义的工作步骤正逐步淡出我们的专业视野，而手绘图这一景观建筑师必备的基础能力也被年青一代设计师所忽视，这种方向上的偏差以及由此对这个行业未来发展带来的消极影响也是不言而喻的。

我们今天所关注的正是竞图大赛中为官方所忽略的那些柯布西耶式的间断性图纸及其表述语言。其中大体涉及题材的取舍、元素的提炼和个性化的表达方式，以及将各种元素权衡、综合之后形成设计师自己的语言和阶段性成果的表达方式。我认为，这种概括性语言和表达能力的提炼，对于个性设计风格的形成和完善具有深远意义。同样在大规模项目团队协作中，这种能力也可以成倍提高各级交流的效率。

本书所选作品以彩铅和马克笔混合制作为多，极少数早期作品中使用过水彩。因为在强调快速、达意的交流性的图纸中，马克笔、彩铅有其天然的优势，我主张大力推广这种画法。为达到这一目的，我在书中的例释和作品集等部分以文字说明的形式，总结了多年景观绘图的经验，以及一些粗浅易学且相当实用的做法惯例，以期与读者分享，并就快速作图技能的提高给广大学生、考生和初级设计师提供一些线索。

需要说明的是，书中有些极快速条件下的图例，快到10分钟成图。在构图、用线等方面均未加以考量，更类似色彩速写。为的也是适应考研中的快速方案表达的需要。在今年的几期教学录像中，我也做过类似的示范，读者的评价褒贬不一，大致是认识角度差异所造成的。但我想指出的一点是，这种快速训练对准备考研和入职考试的设计师非常有必要。就像美院的

wonderful researching drawings were directly rejected in the government's bidding of "pictures". Unfortunately, this awkward and embarrassing situation in designing field is occurring everyday in contemporary China. Because of the limitation of appreciative intelligence, artistic and technical knowledge, and the drive of economic and political considerations, hand drawing, as a determinative designing process, is fading away from our specialized field. And this indispensable basic skill is also ignored by the young architects. It is an obvious developing deviation and it brings negative effect to the future development of our industry.

So what we are concerning about nowadays is the Le Corbusier-style discontinuous drawing and expression. It includes the choosing of subject matter, the extracting of elements, and the personal way of expression, as well as expressive language and way of expressing phased achievements of the designer's own after considering and integrating various elements. I think the refining of recapitulative language and expressive ability not only plays a very important role in the forming of personal style of design, but also helps to multiply the efficiency of communication among all sectors in large-scale project cooperation.

The works selected in this book are mainly rendered by color pencils and markers (rare early works used gouache), for I strongly recommend taking the natural advantage of color pencils and markers in communicative drawings in which quick and expressive are emphasized. For this purpose, I presented textual description for the illustrations and works in this book. I wish to share my years' experience in landscape drawing and some simple but practical skills with the readers, and give some clues of improving quick sketching skills for the students, examinees and junior designers.

It should be noted that some speed sketch illustrations in this book didn't spare too much consideration on framing or applying of lines and are more of color sketching, but seek a quick expression that can be completed in 10 minutes, which is also in accordance with the demands of graduate admission examination. I also made similar demonstration in some teaching videos this year, but received praise

学生 1 小时交 10 张"高速"速写图一样，实质是要求设计师瞬间把握主旨，既是练手头也是练眼力，对设计水平的提高是大有裨益的。这一过程应该尽量快捷，而且绘制过程应该不拘一格，可以使用各种各样触手可得的材料，快速成图，可以快得像天书，像安藤那样，也可以稍微细致，以便于团队交流或作为与甲方交流的图像参考，但作为工作草图的设计表达总的原则是快！这种思考性创作有时会快到难以想象，由于技巧纯熟，作者得以全神贯注于方案本身，而无须顾及表达技巧，最佳状态是心手合一，设计场景在思考过程中似乎能够自动生成；这种创作往往具有高度概括的真实性，完全摒弃了照相式的真实记录，只记下作者感兴趣的作用于设计思维的那些要点，与设计无关的细节一概省略，所以它才既快又"准"（切题），而且更有艺术感染力。这是电脑图纸无法完成的阶段性任务，也是对设计起决定性作用的一步。

我们不妨回溯一下设计手绘图的基本目标——工作交流、解释方案、抓取灵感。我的观点是，草图要快中求好，而非反之。既然你能在 10 分钟之内，用寥寥数笔就表达出场地的主要特征和主要的设计意图，那有什么理由为此花上 10 小时甚至一星期时间去做完美的渲染，以至于当你终于完成那些"完美"的渲染之后，设计的激情、灵感就消失殆尽，这种繁复的长时间渲染在今天完全可以交给电脑来完成。

当前，景观建筑手绘教育和训练受到了前所未有的重视，各种论坛和全国性的竞图活动也相当频繁，但就我的观察，此类活动仍存在一些认识上的偏差，包括对作品的评价更多的还是关注于图面的最终效果，对于设计阐述以及快速高效地表达设计意图等方面的内容关注明显不足，大赛的获奖者和参与者也以工艺美术以及环境艺术专业的学生、教

and depreciating mixed feedback from the audience due to different points of view. Nonetheless, I wish to point out the necessity of quick skill training for the exam for postgraduate and designer applicants. Just like art school students are asked to hand in more than 10 "speed" sketches in an hour, this requirement, essentially, is to seize the subject in a minute by insight and drafting skill training, which helps to improve the design. So this process should be done as quickly as possible yet the operation can be as free as possible: use any materials available to finish a draft. No matter too hasty of slightly delicate as Ando's works for better inside communication or outside reference, the general principle of this sketch as design presentation is to be "quick"! Sometimes this creation from thinking is too quick to imagine, but a skilled illustrator would devote himself into the schema and ignore the expressive skills, and almost automatically draft out the designed scenery in his mind when in a perfect state. Because of its highly generalized truths, this kind of sketches completely abandoned photographic record of reality. Instead, it takes down what the illustrator is interested while ignoring all the details that are unrelated with the design. Being so quick and "accurate" and with stronger art appeal, hand drawings play a determinative role in the design, and this phased task can never be done by computers.

We might as well take a review of the basic aims of hand drawing in design: to communicate, explain, and capture inspirations. I think good is the sub-pursue on the basis of speed, but not the contrary. Since you can express out the major characteristics and design intention with a few strokes in 10 minutes, why be bothered to spend over 10 hours even a week for a perfect rendering? Very likely that your passion as well as inspiration would be exhausted after you finally finish those perfect renderings, so do leave the trivial and tedious rendering tasks to the computer.

Now the education and training of hand drawing for landscape architecture has drawn more attention than anytime before, and various seminars and nationwide competitions are thriving in the field. However, from my

师为主，高水平、大尺度的城市设计及景观综合表达方面的内容涉及尤其少，这与国外许多一流景观设计机构的情况形成了极大反差，所以本书在篇首就开宗明义用不少的笔墨探讨了手绘效果图在当前景观设计中的必要性，以及更有效地利用手绘技能服务于景观规划设计的种种训练方法。并致力于用作者亲自设计完成的大量设计项目为例，从细节片段到作图全过程展示，详细介绍了快速景观手绘图的表达方法和技巧，以期对提高当代设计师的手绘技艺有所帮助。同时，这些直接来自于实际工程项目的大量手绘图纸，也向读者展示了景观手绘图在当代景观设计项目中广阔的应用天地。

高度成熟和个性化的手绘图在表现效率、产品艺术性等方面毫不逊色于电脑制作。在诸如意境表达、多种意象展示等方面比电脑渲图更快捷、更切题。目前国外的许多一流景观设计团队中，手工渲图仍是得到普遍认可和高度成熟的表达途径之一。我们因为认识和教育等方面的偏差，曾一度荒废的手工渲染技术，有必要得到重视和一定程度的恢复，这也是撰写本书的目的之一。

observation, this kind of activities still has derivation in approaching: the evaluation of works still focuses on the final visual effect, obviously insufficient of concerning on design presentation or the quick and efficient expression of design intention; the winners and participates are mainly students or teachers of art school or environmental art; the outstanding large-scale urban design and comprehensive landscape presentation are rarely seen, which indicate a huge gap between Chinese designers and many foreign first-class landscape design studios. So this book discussed the indispensable role hand drawing plays in contemporary landscape design at very beginning in more words than necessary, followed by hand drawing training skills for more efficiently serving landscape planning and design. All the illustrations in this book come from my own project designs. I try to demonstrate expressive methods and skills of quick landscape hand drawing in every detail and whole process, wishing it would help improving the hand drawing technique of contemporary designers. Meanwhile, I believe those hand drawings which are directly drawn from real projects can certainly present readers a larger stage for the application of landscape hand drawing in contemporary landscape design projects.

A highly matured and personal hand drawing is not inferior to computer drawing in expressive efficiency and product artistry, and even faster, more convenient and more accurate than computer rendering in expressing mood or certain imagery. Nowadays, hand rendering is still a universally-received and highly-matured way of expressing among many top foreign landscape teams. Also I hope the once abandoned hand rendering techniques shall be noted and renewed from the deviation in recognition and education, and that is another purpose of this book.

2 景观元素的表达
Chapter II Expression of Landscape Elements

2.1 树的表达
2.1 Expression of Trees

单棵平面树的画法

一个树圈的表达：画出栽植点、轮廓和针阔叶之区分。就表现效果而言，还有分枝、分叶和投影的变化，以及色彩、季相的不同。本节列举数例彩铅和油性马克笔的混合表现，重点在树形变化的练习与记忆。一般掌握3～5种左右的树形变化足以应对平时的作图需要。

首先，树圈用较为浅的颜色上第一层，底色来定色彩的基调和大的明暗关系，常会有些留白来表现受光面。然后，用固有色加笔触来表现树的特征，区分出树的种类。最后，根据环境和光线来增加颜色，让树更加饱满。画投影是最后一步但也是非常重要的一步，不仅区分了层次，而且使整体空间关系也一目了然。

常绿植物需要用冷色或暖色来调配，使其产生丰富的变化，绿色包含黄蓝两个色素，黄色可以有柠檬黄、中黄、土黄、褐等不同冷暖变化，蓝色也是如此。甚至可以用淡紫、群青、赭石来表现其受光的不同感受。

阔叶
Broad-leaf

针叶
Needle-leaf

Drawing Single Plane Tree

How to express a tree circle: distinguish the plantation, contour and difference between needle- and broad-leaves; in terms of presentation effect, the changes of branches, leaves and projections as well as color and seasonal variations shall also be considered. In this section, examples where color pencils and permanent markers are combined to mainly present the changes of tree forms are listed for exercise and memory. Generally speaking, familiarity with 3 to 5 tree forms will be sufficient for the usual drawing.

First of all, draw the first gradation of tree circles with light color. The bottom color sets the tone of colors and main light-and-shade relationship, often left with some blanks to present the illuminated faces. Then, add strokes of proper colors to present the tree's features so that the species is shown. Finally, add colors according to the environment and lights to make the tree appears more vivid. Drawing projections is the final yet very important step, not only gradating the picture, but also making clear the overall spatial relationships at a glance.

Evergreens shall be presented by mixing cool or warm colors for variations. Green contains yellow and blue pigments; yellow can change between warm and cold in lemon yellow, medium yellow, yellowish brown, brown and others, so does blue. Even lavender, ultramarine blue and ocher can be used to present different illumination effects.

此外，色彩搭配宜活。事实上，大比例植物平面极易画"空"、画"板"，尤其不能简单处理，更不能以一种绿色表现所有层次，植物并不都是绿色。1:500 以下的小尺度、大比例的平面图表现形式，细节更易丰富。成组植物平面重在乔、灌、草三层搭配，包括明度、冷暖等对比；其次，投影的勾勒同样有乔、灌、草之分，前者厚（代表高大乔木），后者薄（代表低矮灌木），草皮用点状、排线两种形式，赋色的同时，用笔触分出等高线。

Color matching shall be flexible. In fact, large-scale plant plans often fall into "emptiness" and "rigidness". Simple processing shall be avoided, not to mention representing all gradations with one single green color. Not all plants are green. Small-dimension, large-scale plans under 1:500 are much easier to present abundant details. Plant group plans should highlight the mixing of three gradations, namely trees, shrubs and grass, which involves brightness and warm-and-cold contrasts; projections outlines should also take in account the distinctions between trees, shrubs and grass. With foreground "thick"(representing tall trees) and background "thin" (representing short shrubs), and turfs in dots and rows of lines. Contours shall be made by strokes at the time of coloring.

（1）大乔木体量较大，一般处理为球形（部分留白）。

（2）更大的主干树可以只表现大体轮廓和投影，重点表现树下叠压的空间，如下的组合图，均留一半空间表现上层大树，剩下一半空间表现灌木丛。

(1) Tall trees have large volume and are generally presented as spheres (with partly left blank).

(2) Larger mainline trees can be presented through general outlines and projections only, with graded spaces under the tree highlighted. In the three combined drawing examples below, half of the space is used to present the tree above, and the other half is for shrubs.

从构图上反映出植物层次和群落特征，三五成群、多层叠压、相互掩映是植物生长的基本规律，也是符合美感的平面构成形式，这里除了大小、层次变化之外，色彩的冷暖变化显得尤为重要，关键在于大胆用色，小心收笔，常备高对比度的两组马克笔和一支灰色马克笔。大乔木用色温和，小灌木可以用非常跳跃的色彩点出生机。最后的黑色投影是所有层次的调和色，阴影使画面扎实，阴影让植物"落地"。平面图纸的清晰与层次区分全在阴影的勾勒。

层次的区分，叠压全在阴影勾勒（黑色）；在暗底上用细线涂改液加画一层（白色）。

The gradations of plants and features of plant communities should be reflected in the composition of picture. Plants grow in threes and fours, with multiple gradations folded together and setting off one another, forming planes pleasing to our eyes. In addition to the variations of size and gradations, the change of colors between warm and cold is particularly important. The key of presentation is using colors boldly and being careful in tidy-up, with two sets of markers of highly-contrasted colors and a gray marker always at hand. Big trees should be in soft colors, while small shrubs can be presented by bouncing colors to create a sense of vitality. The final projection in black is for harmonizing all gradations. Shadows make the picture solid, and the plants "landing". They are the key to make the plan clear and distinguish gradations on it.

The key to distinguish gradations, therefore, lies in the drawing of shadows (in black); on the dark ground color one more gradation (in white) is added with fine line white-out.

景观表现从树开始，本书开篇特别从作者以往速写册里标出数十幅画树草图，从一颗点景大树开始，由真实之树、画家之树，逐步向意念之树、地域之树过渡。画法兼顾了设计师常用的线笔、铅笔、炭笔等多种技法，画"骨"为主，基本不涉及着色技巧，目的是让读者更清晰了解树形、出枝等方面的基本用笔；用细线和墨块两个层次表现，简单易行，开笔有益。

　　从深到浅，由粗到细。用弹性的细线表现主干的质感，主干根部及树冠阴影用墨块"点"出，虚实轻重、缓急、光与阴等因素均在墨块之"点"。

　　Landscape presentation starts from trees. At the beginning of this book, tens of pictures of trees and grass are displayed, all chosen from the author's sketch books in the past. The first is a landscaping tree, showing the techniques to graduate from real trees and painters' tree to the tree of ideas and territories. Multiple skills are used, liners, pencils, charcoal pencils and alike. "Bone" is the main object of drawing, which seldom involves coloring techniques. In this way, I hope to give the readers a clearer understanding of basic strokes in presenting tree forms, branching and so on; the trees are presented through two gradations, namely fine lines and ink blocks, simple yet useful to practice.

　　From dark to light, and from thick to thin. Elastic fine lines are used to present the texture of the trunk, and trunk root as well as crown shadows are "dotted" through ink blocks. Virtual and solid, light and weight, slow and hasty, light and shade... all these elements appear in form of the "dots" of ink blocks.

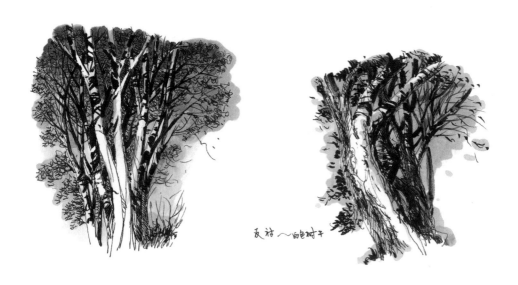

常见的大树表现

（1）大树一棵为主足矣，表现上注意：有脚（落地，近人），无头（高大，留下想象余地）。

（2）出枝以一侧为主，切忌伸臂布指、左右均衡。

（3）主干多为"衬出"而往往不需要"死抠"。

（4）出枝之处，均有阴影——是最出彩之处，宜用力刻画。

Usual presentation of trees

(1) Take only one tree as the main body for presentation, note that it should: have root (falling to the ground and close to people), be headless (tall, leaving room for imagination).

(2) Branches on one side shall be mainly presented; do not extend too many branches and leaves, or attempt to achieve balance between both sides.

(3) The trunk is always presented through "setting off" instead of "cutting out".

(4) All branches shall have shadows, which is the most appealing element and the strokes should be solid and firm.

孤植大树一般体量很大，表现稍不充分就会显空；层次稍多又会出现杂乱、花、脏等情况。我的经验是，出枝顺方向（一侧出枝），明暗成体系（2～3层），力求删繁就简。一般而言，主干须多留意质感，上部枝条则要留意穿插，前后之间的剪影和深色叶片相衬托。

孤植的树出枝要表现出前后关系，一组之中也要有前后，一般以两三树一组为宜。

Isolated trees are often very huge, making the picture vacant if not sufficiently presented; however, if there are too many gradations, the picture becomes cluttered, disordered and dirty. To my personal experience, we should make it simple by removing all complications. Branches should mainly come from one side, and light and shade should be presented systematically (in two or three gradations). In general, for the trunk the key is texture, and for upper branches intercrossing; silhouettes and dark leaves in the front and rear should set off each other.

On isolated planted trees, branches in the front shall be distinguished from those behind. So shall trees in a group, usually of two or three trees.

突出光影的表达需利用有韧性的粗水笔线条直接"扫"出重要的枝条，突出主干"双勾"，主干的表达完全依靠上部枝叶的投影。同时，通过前疏后密的线条排布衬托出前后三个层次，即空白的主干、乱线衬出的树叶和粗黑线条的背景枝干。

In an expression where light and shadow is highlighted, major branches are drawn by "sweeping" resilient thick lines with fountain pen, so as to emphasize the trunk's "double hook". The trunk is entirely expressed through the projections of the upper branches and leaves. Meanwhile, the lines, arranged sparse in the front and dense behind, create three gradations, that is, the blank trunk, leaves appear through contrast and the background branches drawn by black bold lines.

北方的特色落叶松，主干如铁，结构感强烈，枝叶部分则柔软如发梢，形成团簇状的质感，同样做出前疏后密的层次，枝叶的点化需要多次磨炼，也可以用粗线条的笔触表达类似的效果。

Larch as a featured species in North China, their trunks are strong like iron structures yet branches and leaves as soft as hair tips, forming a texture of clusters. Expression of larches also requires grading so that they seem sparse in the front and dense behind; many practices will be needed to get familiar with the dotting of branches and leaves, while bold strokes can also express the similar effect.

柏树在中国古建景观中，常出以雌雄双株，一瘦弱、一壮硕，特征明显，前者叶衬干，后者以雄壮有力之感为前景，主干表现宜繁华，分枝则多虬曲劲健，树瘤树节是枝干表现的又一重点，尤其是中国式柏，此为一大审美趣味。其表现同样需要有主次筛选，形态上有意分出正、侧、俯、仰的变化。

中山公园北口背映红墙，沿河多此千年古柏，辽代肇建南京之时即以此地为京都苑囿。图中巨柏历八百年留存至今，世之柏树，古拙奇崛无出其右，古华百年，为之写照传神者代有其人，又岂独画家哉。

古柏细节——树疣表达，全在线条之疏密变化
Ancient cypress details — line spacing changes is the key to the expression of tree warts

In Chinese ancient architectural landscapes, cypresses often appear in pairs, one male and one female, with the obvious feature that one is slim and weak and the other bulky and strong. The female tree has more leaves that sets off the trunk, and the male has powerful trunk and branches as the foreground. The trunks should be presented in a prosperous manner, and branches curling and strong. Knobs and knots, as an major aesthetic element, should also be valued, especially for Chinese cypresses. They should also be divided into primary ones to be highlighted and less important ones, and represented in different forms, namely obverse, side, pronating and upward-facing.

The north entrance of Zhongshan Park backs red walls, with numerous such ancient cypresses along the river. It had been a capital garden since the South Capital was built in the Liao Dynasty. The giant cypress has undergone 800 years and survived to now, with primitive simplicity and marvelous beauty unparalleled by any other cypress in the world. During the hundreds of years, not only painters, but many others had tried to depict its image and charming

以线为主的表达：中国传统的古柏多重枝干表现，一树之精神全在树干的虬曲穿插，基本构图不谬之外，全在排线练习。

Line-dominated expression: in Chinese traditions, branches are the focus in expression of ancient cypresses. In other words, the spirit of a cypress lies totally in the curling and intercrossing of its trunk and branches. Therefore, the painter, apart from ensuring the basic composition of picture is free of error, should put all efforts in doing line practices.

线面结合的表达：苍碧天跷之古柏，宋代太湖石座，衬出故宫金色琉璃之一角，是为中山公园之典型景观。

Expression of lines and planes: ancient cypresses under the pale green sky, together with the stone seat of the Song Dynasty, set off a corner of a golden glaze of the Forbidden City. A typical scene of Zhongshan Park.

树干的细节及木材的质感描绘
Detailed dipiction of texture,fabric of wood materials

杂树林的表达：一片杂树林往往充斥着各种树木，落叶乔木，或者常绿松柏。画好一片杂树林最重要的是前后层次的表现，如下图，前面几乎留白的大乔木和后面浓重的松林无论在色彩深浅还是在造型处理上，都有着强烈的对比。

炭笔的运用：炭笔线条柔软，容易画出细腻的质感和顺畅的过渡，用色块的轻重扫出树林的层次，再以细而浓重的线条画出主干的前后穿插，树叶多集中于树梢和上部树冠，宜用柔软的灰色块面涂出，使之成团簇状分布。

Expression of mixed forests: a mixed forest is often filled with a variety of trees, deciduous ones or evergreen pines and cypresses. The most important thing in drawing a mixed forest is properly to express the gradations in the front and behind. As shown in the below drawing, the big trees almost left blank in the front is in sharp contrast with the dense pine trees behind them, in both color shades and form presentation.

Use of charcoal pencils: good at soft lines, and suitable for presenting a delicate texture and smooth gradation. First produce the gradations of woods through "sweeping" color blocks of different shades, and then present the intercrossing of trunks fore and aft with thin, thick lines. Leaves, which are mostly distributed in the treetops and upper canopies, should be presented in grey blocks and faces to shown the clustering form.

炭笔表现的树桩：树桩本身笔触较少但要求造型精准，树桩背景的层次如灌木小枝反而笔触较多，进而衬托树桩使其靠前，拉开层次。

彩铅线条的表现：单色彩铅表现树丛能够达到"复古"的效果。尤其是水溶性彩铅，附着力强，通过用笔轻重能够画出较多层次的色彩。

用彩铅画一片树林时，不求面面俱到，靠前的树要刻画得深入一些，后面的层次逐渐弱化，用较深的树干、出枝衬托出较浅的（甚至是留白的）团块状的树叶。

Tree stump expressed by charcoal pencils: not many but precise-formed strokes are used for the stump itself, while background gradations such as the shrub branchlets have more strokes to "push forward" the stump by showing different gradations.

Presentation by color pencil lines: single-color color pencils can be used to express woods for a "vintage" effect. Particularly water soluble color pencils, because of its strong adhesion, are able to present multi-gradation colors through different strengths of laying strokes.

When using the color pencils to draw a wood, do not attach the same importance to all aspects. Depict trees in the front more thoroughly, and gradations behind it should be weakened one by one. Leaves in the shape of blocks with lighter colors (and even left blank) should be set off by trunks and branches of deeper colors.

罗马梯沃利的艾期特花园写生——花园里的柏树
Sketch for Villa d' Este of Tivoli, Rome

省略了部分细节，增加了前后层次，从自然丛林的描绘逐步转向人工造景因素的多层次景观图像。

Some details are skipped, and gradations from front to back are increased. The depiction of natural jungle is gradually changing to multi-gradation landscape images with artificial landscaping elements.

事实上，大量的风景区、植物园和公园的规划意向表达与传统意义上的风景画几乎是无法区分的，最早的职业风景设计师也多为风景画家出身，二者的联系性是毋庸置疑的。也正因如此，类似《红皮（Red Book）式书》的风景构造尝试至今仍被认为是有一定指导意义的。熟练的设计师可以在极短的时间里将这种"速写式"风景转化成极完美的风景设计图。远山、丛树、中央场地和近景大树共同构成的风景之美，200余年来并没有什么本质的不同。今日的设计师仍有必要向我们的前辈一样，多用手眼，而不是一味依赖深不可测的所谓"理念"去实现所谓"设计"的专业性。

In fact, a lot of planning intention expression of scenic spots, botanical gardens and parks can be hardly distinguished from landscape paintings in the traditional sense. Most of the earliest professional landscape designers had been landscape painters at first. No doubt the two are closely connected with each other. Exactly for this reason, landscape construction attempts similar to the Red Book are considered as instructive in some sense to this day. Skilled designers can, in a very short period of time, convert such "sketched" landscape into perfect landscape design drawings. The remote mountain, clump of trees, the field in the center and close-range trees constitute a scenery whose beauty has never changed essentially across more than 200 years. For our designers today, it is still necessary to, like our predecessors, use their hands and eyes, rather than achieving the so-called "professional" design depending on nothing else but the unfathomable so-called "concepts".

钢笔线条的表现：钢笔线条不易修改，但细致的钢笔线条密布能表现出极为细腻的光影变化与层次对比，笔者习惯以细钢笔与一支弹性粗毛笔配合，以钢笔密排出树林的灰调子，再以肯定的粗笔"点"出光影与暗部，色彩层次明确，是简单易行的练习方法，读者可以一试。

Presentation by pen lines: pen lines are hard to modify; they are able to present the finest light and shadow changes as well as gradation contrast through delicate and densely arranged lines. I am used to use fine pens and a flexible writing brush. With the pen the grey tone of the woods is presented with dense lines, and light and shadows by "dotting" assured thick strokes, so as to present clear color gradations. This is a simple and operable way for practicing and can be referred to by my readers.

切特伍兹花园速写，由松和橡树丛组成的布朗式风景。连绵的缓坡草地构成英国式牧场风光的基底，布朗式的点景树一般为苏格兰红松或橡树构成，日间阳光下的影子随着缓坡和丛林的分布不断变化，气质温润与地中海一带强烈日照下的阴影形成极大反差。英伦特有的晦暗湿润的光在布朗的园林中用橡树的影子体现，用布朗的语言，就像是丝绸一般滑过山坡的温柔阴影。

当年的风景式花园是依据马车的高度和速度设计的竖向与尺度，今日的访客与风景的描绘者应深谙此道，布朗先生希望我们站在高坡上去欣赏隔着蜿蜒风格的小溪流，去欣赏对岸橡树的缓坡和苏格兰巨大的红松孤植树。这时候，红松犹如加州约瑟米蒂的红豆杉林一样，会凸显出高大雄伟的气质。

Sketch of Chatsworth Farm and Cottage, a Brown-style landscape is made up of pines and oak trees. The continuous and gently-sloped grassland forms the substrate of the British-style pastoral scenery; Scots pine or oak trees are often used as the Brown-style landscaping trees; in the daytime, shadows keep changing as the gentle slope and jungle distribution varies; the mild temperament is in great contrast to the shadows under the strong sunlight in the Mediterranean area. In Brown's garden, Britain's unique dark and moist lights are presented with shadows. In Brown's words, they are soft shades that slips over the hillsides like silk.

Vertical dimensions of the landscape garden of that time were designed to suit the carriages' height and speed. Today's visitors and scenery depicters should be very familiar with this. Mr. Brown wanted us to stand on the high slope to have a look at the small meandering streams, to see the gentle slope of oak trees on the opposite bank and the big isolated Scots pines. At this time, the red pines stand out in a lofty and imposing gesture like the yew woods in Yosemlte National Park, California.

巴黎卢森堡花园入口
Entrance to Jardin du Luxembourg, Paris

默伦的维康府邸侧门入口景观
Landscape at the side entrance to Vaux-le-Vicomte, Melun

单体与组合树的表达：（1）很多时候树干是可以空开留白的，无论上色与否，都更宜于表现。（2）夏季时草木葳蕤，密度较大的乔木可以采取概念性的画法。那么如何去画呢？要注意树叶是连片的，运笔方向要一致。密中有空，透出天光来，否则易死板。

Representation of single tree and the mixed trees: (1) for a lot most of the time the trunks can be left blank; whether colored or not, they are much easier to be presented; (2) in the summer when vegetation is luxuriant, dense trees can be presented through conceptual techniques. Then how to paint them specifically? Note that leaves are in blocks, the direction of brush strokes must be consistent. Among the thickness there are gaps where daylight shoots down, getting the picture out of artificiality.

杂树林的速写：采用具有一定弹性的水笔画细线，用勾线毛笔画粗线和点叶，形成粗细的层次，完全单色速写，类似木刻版画效果，用笔须肯定，突出光影效果。

Sketch of mixed forest: fountain pen with certain flexibility is used to draw up fine lines, and brushes for heavy lines and dotted leaves, so as to present grading of thick and fine lines. It is an entirely single-colored sketch with an effect similar to woodcut print. The strokes shall be firm to highlight the light and shadow effects.

用油性马克笔直接"刷"出柏树的质感，油性马克笔在具有一定渗透力的厚纸上可以表现出色彩差别极大的色彩变化。一色用笔，关键是运笔速度变化，横扫出飞白与浓淡的多种变化，配合长长的投影，非常适合表现秋日艳阳下的植物景观与色相。

Permanent markers are used to present the texture of cypresses via direct "brushing". On thick papers with certain penetrability permanent markers can express great color changes. The key to use one-color pens is the speed changes of brushing. Sweeping-up at different speeds results in hollow strokes or varied color thickness, quite suitable for expressing the plant landscape and hues under the bright sun in the autumn.

柏树的基本形态：中国园林古柏表达多属上实下虚，树干形如发丝婉转，下疏上密的树冠形成浓密的结顶。与西方园林中的笔柏表现形态正好相反，后者随光影变化，上虚下实，在树下形成浓密的束状阴影（本页为笔者一堂普通的作图课所做的黑白小稿，共约 1 小时完成）。

Basic forms of cypresses: ancient cypresses in Chinese gardens are mostly solid in the upper part and hollow in the lower; the trunks are in the shape like gently wound hairlines, and the crowns, sparse downside and thick upside, form dense apexes. This is exactly the opposite to the manifestation of pencil cedars in Western gardens, which, varying with light and shadows, are often hollow in the upper part and solid in the lower, producing thick shade in bundles under the tree (below is a small black-and-white work of the author in an ordinary drawing class, completed in about an hour).

树有向北、粗细、雌雄之分，无论俯仰姿态如何夸张，皆须有相互呼应之势。

Trees are distinguished by their gestures (obverse or reverse), size (thick or thin), and gender (male or female). No matter how exaggerated their gestures are, they shall echo each other.

树与阳光，投影的变化与强度是表现的趣味中心。

The major interests of presentation fall into trees and sunlight, as well as changes and strengths of projections.

树的立面在场景中的表达：投影的疏密和透视变化指出了空间的延伸方向，具有规整形态的笔柏是这类透视的最佳表现题材。

Expression of trees' facades in a scene: the density and perspective of projections indicate the direction of spatial extension; pencil cedars, with clear and regular forms, is one of the best themes for such perspective.

白桦林的投影方向表现出地形

The projecting direction of birch woods shows the landform

2.2 天空的表达
2.2 Expression of the Sky

天空的用色通常是逐渐过渡的：

（1）色彩是有情绪的。（2）颜色由深浅过渡，通常上深下浅，在接近地平线的部分露出一抹亮色。（3）整体天空退晕要"抹"，要抹得均匀，亮色用彩铅隔出来，块状用笔要肯定，不能来回抹。

右图实际划了片天空：A、B、C、D 各自可以独立成完整的天空。从阴到晴，从远到近总会有一抹亮色提示阳光的存在，黄昏的云彩之下常常会有一带亮色。像美国人常说的："Every cloud has a silver lining"，既是自然真相，也是人生哲理，我喜欢这一抹银边，常常用涂改液专门细细地刻画出这一幕。

Coloring of the sky is usually a gradual transition:

(1) Colors have emotions. (2) The sky, with color shade gradation, is often of deep color upside and light downside, with a splash of bright color near the horizon. (3) The color grading of the whole sky is realized through "wiping", which shall be even, with bright colors isolated with color pencils, and color blocks presented with solid stokes without wiping back and forth.

In the right drawing, the sky is actually divided into different parts, namely A, B, C and D, which can be completed independently. From cloudy to sunny, from far to near, there will always be a touch of bright color that remind us of the sunlight. It is often presented below the clouds in dusk. As the Americans always say, "Every cloud has a silver lining", which is not only the truth of nature, but also the philosophy of life. I like this touch of silver, and often carefully depict this scene with white-out.

水与天空的色彩关系：水面的色彩几乎就是天空的再现，当然是艺术化的再现而绝不可能如镜子中世界一般亦步亦趋。右图是专门绘制的热带海滨，同一场景、统一大范围的海天一色的表达，由色彩选择的差异反映了不同的感觉、不同的天气。

场景1：早晨的阳光穿云而出，天空基本湛蓝，水的远景是天空上下色彩的中和，近景则是海水与沙漠色彩的中和，水面平静深远。近景的礁石激起浪花。

场景2：夕阳下的绚烂是主题，一层层的海浪由远及近，迎光面映出天空绯红色的晚霞，背光面则是海水本色，交替进行，直至近景与沙滩的黄色相混。沙滩的色彩也并非简单的一抹黄色，一般由远到近也会呈现出海波之蓝和近岸沙丘的涂色，当然这些变化都应在整体协调的色调控制之下，免于单调，但也要防止色彩过花哨流于纷乱。近景总是需要一抹重色以"压住"画面，海岸松是较好的选择。松树的笔法宜简约、有力，在画树的篇章中有多个图例说明，此不赘述。

Color relationship between water and the sky: the color of water surface is almost a reproduction of the sky, nevertheless an artistic representation rather than an exact copy like the image in a mirror. Right are specially painted drawings of a tropical beach. With the same scenario as the subject, both expressions of a wide range of sea and sky in the same color, they demonstrate different feelings yet both being vast and magnificent.

Scene 1: the morning sunlight penetrates through the clouds, the sky blue on the whole; the remote part of water is a neutralization of upper and lower colors of the sky, and the close part a neutralization of the seawater color and desert color, with the water surface calm and deep. Reefs in the foreground arouse waves.

Scene 2: The gorgeous and dazzling sunset glow is the theme. Layers of waves come from far to near, with the scarlet sunset clouds reflected on the illuminated area. The back facet is in the natural color of seawater. These colors transit alternatively till they are mixed with the yellow sand in the foreground. The sand beach, however, is not simply in yellow. It also take on the blue of sea-waves and color of sand dunes near the shore from far to near. Of course, these changes shall be subject to the overall color tone control and coordination to avoid tediousness, but the use of too fancy colors shall also be prevented, which will cause intricacy and disorder. A touch of heavy color is always needed in the foreground to "hold" the picture, and the maritime pine is ideal for this purpose. The pine should be presented in simple yet powerful strokes, which is demonstrated through a number of figure examples and therefore not repeated here.

富士箱根河口湖的天空：黄昏远山森林构成远景层次，近景的湖水映出天空的色彩；毫无疑问这里的天空也是过渡的，高空的色彩总是冷蓝色，低空总是绚烂的金色。远处的白头富士山，近处的日本鸢尾花构成画面地域性景观元素。

Sky of Lake Kawaguchi, Hakone, Fuji: the remote mountains and forests in the sunset constitute a distant view, and in the foreground the lake water reflects the color of the sky; no doubt the sky here is gradating in colors, with cold blue in the upper air and gorgeous gold at lower airspace. Mount Fuji in the distance with its top covered by snow, as well as Japanese iris flowers together are the regional landscape elements in the picture.

Innsbruck vista.
—WJ 2014.1

马年元月,由南德至奥地利沿途皆漫天大雪,至此因斯河畔小城,天竟放晴.借黄昏夜色写城北小山村.距此不远处即有扎哈设计之滑雪跳台.助输小记于深dll

    马年元月，由南德至奥地利沿途皆漫天大雪，至此茵斯河畔小城，天竟放晴。借黄昏夜色写城北小山村，距此不远处即有扎哈设计的滑雪跳台。

    最直接的画法：用色彩平推，逐层退晕。用白色涂出卷云，暖色集中于云的下层，水面要与之保持一致。这一类作品几乎无法区分景观图与风景画之界限，唯一的区别只在于用色的厚重与层次细腻之别。

    In the first month of the Chinese year of horse, snow fell thick and fast along the way from in South Germany to Austria; unexpectedly it became clear when we arrived at this small town on the bank of the Enns River. By the dim light in the dusk I sketched the village to the north of the town, not far from which is a ski jump designed by Zaha Hadid.

    The most direct presentation: "push flat" with colors for gradated color change. Draw up the cirrus clouds by "daubing" the white paint; warm colors concentrate at the lower layer of clouds, and the water surface shall be the same color. Such works can hardly be classified as a landscape plan or a landscape painting, as the only difference lies in thickness of colors and delicacy of gradations.

地中海天空、土地、晚霞中的乱云。山雨欲来，霞光一线，乱云连片，要块状用笔，笔触肯定，空出的飞白须自然。翻滚的云层，形态其实和波涛汹涌的水面十分相像。要注意表现天空不光只用蓝色，还可以添加杂色来表现，正如我们在库尔贝或柯罗的风景画中所见一样。

Sky, land, and scattered clouds in the sunset of Mediterranean. Mountain rain is coming, with a ray of sunlight, and scattered clouds are connected with each other into pieces. Use firm strokes, with hollow strokes being natural to present the blanks. Rolling clouds are much like the turbulent water surface in the form. Note that more than blue, mixed colors can be added to present the sky, as we've seen in the landscape paintings of Camille Corot or Gustave Courbet.

乱云中的阴与晴，笔触！笔触！把握不足时就稍稍多画几笔，没关系，记住喽！

只有半张 A4 大小的快速马克笔练习，对于大笔触摆放、气氛渲染都有帮助。这种小稿往往不是一次能成功，必须反复磨炼，一旦画成，所有的笔法都将成为以后画图的招牌动作，可以终身不忘。反复磨炼，至为重要！

Rain and shine in the scattered clouds. Strokes! Strokes! Lay more strokes if you are not so sure. Keep it in mind! You are not a John Singer Sargent, so a little verbosity is not big deal. After all, imagery is the girl in your dreams!

Fast marker exercises in the size of only half an A4 sheet is very helpful to stroke arrangement and atmosphere play-up. Such small works are seldom accomplished in one action. Repeated practicing will be required. Once being grasped, these techniques will become your signature moves that never got lost. Repeated practices is the most important thing!

2.3 水的表达
2.3 Expression of Water

　　天空和水面在景观图中所占面积较大，往往对画面的整体效果如色相、空间氛围等起到主导作用。其共同的特征是明快、流畅，流云飞逝，流水不舍。关键在一个"活"字，用笔须活而不乱。故天空整体用色宜简洁、明快，多出以大笔触，多用"刷"、"扫"，无须刻画过死。

　　天空多用淡色马克笔扫出，对于大面积天空的色彩渐变，最好采用饱满的油性马克笔混合同色系的彩铅作退晕，由于光线折射的原因，天空总体用色由高空到近地面处，逐渐由深变浅，由冷变暖。

　　Sky and water, which occupies large parts of the landscape drawings, often dominate the overall effects of the picture in hues, atmosphere of the space, and so on. Both of them feature liveliness and fluency, as clouds flies swiftly and water sheds continuously. "Flexibility" is the key; that is, strokes must be flexible yet in order. So the overall colors in the sky should be concise and bright, with "brushing" and "sweeping" strokes largely used and rigid ones avoided.

　　The sky are mostly presented through "sweeping" with light-color markers. To present the gradual change of colors in large areas in the sky, better use permanent markers together with color pencils of the same color family for color gradating. To show the effect of light refraction, the entire sky shall be colored deeper, cooler in the upper air and lighter, warmer as it gets close to the ground.

暖色调的叠水溪瀑练习，坚挺的块状用笔，阴影与实体皆十分肯定，水面用色柔美灵动，形成对比。叠水用笔以白当黑，用笔宜少而精，飞白不够之处，方用白粉添加。

Exercise of cascading river waterfall in warm colors. With firm strokes in blocks, both shadows and substantial objects are presented in an assured manner. Water is in soft yet vibrant colors, posing a beauty of contrast. In the drawing of cascading water black is composed by white; wherever hollow strokes are not sufficient white powder pigments can be added for make-up.

①定色由环境色和固有色彩。
②质感 流动的水
   坚实的石。
③色温 冷/暖

贵州黄果树瀑布景观规划过程图：

（1）线稿要简洁、轻快，只要表现出最大的结构变化即可，细节不宜多，不要把画面"画黑"，硬质部分可加入排线，以加强结构。

（2）铺大色调，把远景山崖部分色差拉大，容易表达的部分先画，做出大层次，跌水用笔要活，可由浅及深，高光可空出，或用白色涂改液直接画出。

（3）重点刻画近景的山色和溪流，这是本张图的最难处，用色浓重，还要画出水雾和青苔，用三、两重色交替叠加，均以宽笔扫出，一次不行可多次尝试，但不宜来回涂抹，如此便失去了景观师手图的干净利落。此处用马可笔表现实际上比用水彩更难，但一旦练熟，将百试不爽。

Process map for the landscape planning of Huangguoshu Waterfall in Guizhou:

(1) The line work shall be simple and vivid, only to present the major structural change without too many details, and do not put the pictures "painted black". Hard parts can be presented by adding rows of lines to emphasize the structure.

(2) Set the major color tones, and leave distinct differences on the distant mountain edges; present the easy parts first to show the main gradations; apply flexible strokes for the hydraulic drop, better from shallow to deep, where the highlight part can be left empty or directly drawn up with white-out.

(3) Take the mountain sceneries and streams in the close range as the focus of depiction. These are the most difficult part of the drawing, with heavy colors used, and water mist and moss to be presented. Two or three heavy colors are overlaid in wide sweeps; do it several times if it cannot be done at one stroke, but daubing back and forth should be avoid to keep the neatness and tidiness of hand drawings. Actually it is harder to use markers here than to use watercolors, but when you get familiar with it through practicing, it will work well every time and everywhere.

（4）补全画面，调整色调，用浅蓝、水绿等色块，大笔抹出即可。

注意：画面中大量的白色是空出的，而非如建筑鸟瞰中一样用涂改液。此处大量用白，会使画面浑浊，将减损清新爽利的效果。

(4) Complete the picture, and adjust the hue by "daubing" blocks of light blue, light green and so on.

Note: the white areas in the picture is left blank instead of being painted with white-out like the architectural aerial views. Because the wide use of white color here will make the picture cloudy instead of keeping it fresh and neat.

2.4 山石的表达
2.4 Expression of Hills and Rocks

黄果树瀑布景观规划鸟瞰图
Aerial view for the landscape planning of
Huangguoshu Waterfall in Guizhou

黄果树瀑布景观规划鸟瞰图局部
Part of the aerial view for the landscape planning of Huangguoshu Waterfall in Guizhou

贵州六枝山区地貌景观

Geomorphologic landscape rendering, Liuzhi District, Guizhou province

贵州六盘水山区的景观。六盘水六枝特区是典型的喀斯特地貌即两山一谷，山谷中层层的梯田与峡谷石山相呼应，形成有如桃花源一般的感受，中国西南六盘水、攀枝花等城市都有着盆景城市的美誉，从空中鸟瞰整个城市的山水格局有如极度放大的中国园林，城市的所有生产用地位于山脚汇水区域，条条溪流从中穿过，半山为白色的苗族、仡佬族的村寨，犹如一块块的魔方洒在青山绿水之间，从风水格局的角度来看，这也是一个典型的内聚型的"壶"形空间，具有藏风聚气的优越生存条件。

一座山水大盆景式的葡萄庄园。周边的喀斯特石山在全贵州山水当中也算独具特色。在雨水充沛的山区形成了类似于欧洲托斯卡纳区域的农作物风光和庄园景色。唯一的区别是光照条件略显不足，所以二者的农业风光只是在作物品种上形成区别。例如贵州的茶山、托斯卡纳的葡萄酒庄园和南法普罗旺斯薰衣草庄园。作为大地的歌手，应该充分了解这些区域本质性的特征。

Landscape of Liupanshui mountain area, Guizhou. The Liuzhi Special District in Liupanshui has typical Karst landform with two mountain and a valley. The valley, where layers of terraced fields echo the gorge and rock hills, is just like a paradise on earth. In Southwest China, Liupanshui and Panzhihua are reputed as "bonsai cities", with overall landscape patterns looking like extremely amplified Chinese gardens. The land for production is located in the catchment area at the foot of mountains, with brooks flowing through. In the mid-way there are white stockaded villages of the Hmong and Gelo nationalities, scattering among the mountains and waters. From the point of view of Fengshui, this is a typical cohesive "kettle"-shaped space, an advantageous field where *qi* is concentrated and luck gathers.

Landscape in Liupanshui's mountainous area in Guizhou: a vineyard like a huge bonsai. The surrounding Karst rock hills are unique and special among all sceneries in Guizhou. Due to the abundant rainwater in the mountain area, the farm crop scenes and farmland landscape is similar to those in Tuscany, Europe. The only difference is that lighting conditions in Liupanshui is not as good as in Tuscany; so their agricultural landscapes differ only in the varieties of crops. Tea Mountain in Guizhou, vineyard in Tuscany and Provence Lavender Manor in France are such examples. Designers, as singers of the earth, should fully understand the essential features of these regions.

2.5 建筑物的表达
2.5 Expressions of Architecture

　　以下数例快速建筑环境表现图均是最大限度发挥油性马克笔在大范围退晕及快速渲染方面的优势，单幅用时均在 10 分钟以内，建筑色彩与环境背景完全融为一体，建筑固有色彩几乎不作考虑，建筑群组关系完全由倒影、反光、投影加以确认，将建筑的环境特征表现至极致。

　　（1）外轮廓尽量简化，表达出体量，保证透视即可，视点的选择很重要，要考虑建筑的气势。

　　（2）用光影表达层次、空间关系，固有色尽量弱化。

　　（3）背景要烘托亮部，形成反差。

　　In the following examples of rapid presentation drawing of architectural environment, the permanent markers are best used for large range of color grading and fast rendering, so that every drawing can be completed in no more than 10 minutes. The colors of architecture totally fused into the background, with the proper colors of the former hardly taken into account. All relationships among the building complex are presented by way of reverse images, reflective light and projections, bringing into extremely full play of the building's environmental characteristics.

　　(1) The outlines are simplified as much as possible, with the only aim to express the volumes and to ensure perspective effect; the selection of point of sight is very important, which involves the presented momentum of the building.

　　(2) Use light and shadows to express gradations and spatial relationships, with proper colors weakened as much as possible.

　　(3) The background should set off the lighted part to form contrasts.

快速表达示例

(1)光影模式:明确无误的亮区和暗区。

(2)楼群主体色调:或偏冷绿、冷蓝或偏灰褐、暖黄。注意用笔的整体性,干净利落,并适当突出对象的质感,前后三座楼分别表现不同光洁度的立面。

(3)前后景表达:明度上,后景以深色为主,前景为亮色;色度上,后景灰,前景跳跃而纯净;细节上,室外光影,室内灯光相互映衬。城市总是离不开光的辉映,总有一束温柔的光,照进我们的世界。

Expression of Quick Sketch

(1) Lighting mode: clearly-defined bright area and dark area.

(2) Main color tone of buildings: cold green, cold blue and alike, or gray brown, warm yellow and alike. Note that strokes shall be consistent and clear, appropriately highlighting the texture of objects. Different smoothness of the facades of the three buildings in varied distances shall be shown.

(3) Expressing the depth of fields: in terms of brightness, the background is mainly in dark colors and foreground in bright ones; in terms of chromaticity, the background is grey and foreground vivacious and clear; in terms of details, outdoor light and shadows and interior lighting complement each other. Cities cannot do without lighting; there is always a warm beam of light, shooting into our world.

完全由环境色彩控制的建筑，整栋玻璃大楼实质就是环境再现：上部天空，下部楼裙投影，均一一表现，最后加画楼内灯光，形成水晶盒一般的光亮剔透效果。

This is an architecture fully controlled by the environment colors, a glass building that basically reproduces the environment: sky in the upper part and projection of the attached building in the lower part are all presented. Finally interior lighting is added, creating the bright and translucent effect as a crystal case.

第一层线图就开始交代阴影层次
The shading gradation is presented in the very first layer of line work

整体铺陈环境色彩
The overall environment colors are arranged

"点"出城市灯火与辉煌
Brilliant illumination of the city is presented via "dotting"

典型的城市街道空间
Typical urban street space

楼群关系示例

基本规律是上浅下深，逐层加上楼群之间的相互投影，这是最重要的一步，马克笔在刷出挺拔投影和上下渐变时最具优势。

最后的整理，实质是在原有层次基础上刻画出阳光直射在建筑顶部的效果和建筑底部的闪光、街道轮廓等等。上色的核心是建筑之间的投影。

Example of building complex

The basic law is to use light colors in the upper part and dark ones in the lower part, and buildings' projections on each other are added one by one, which is the most important step. Markers does the best in brushing the upstanding projections and up-to-down gradual changes.

The final clear-up work is basically to, based on the original gradations, depict the sunlight beating down the top of buildings, as well as shimmers, profile of streets, and so on. Projection among buildings is the core of coloring.

深圳高铁北站绿谷设计鸟瞰
Bird view of Shenzhen high-speed railway station,"Green Valley"

山东潍坊白浪河滨水景观
Perspective rendering of waterfront area, Bailang River, Weifang City

巴黎拉德芳斯新区景观
Landscape of La Défense new district, Paris

深圳高铁北站绿谷鸟瞰
Bird view of Shenzhen high-speed railway station,"Green Valley"

波士顿港区城市意象

以下一组图表现了波士顿滨水区域的长码头和中央公地公园两个主要场景。这是波士顿区域最具特色的翡翠项链公园系统的组成部分。这种针对典型景观的快速提炼表达训练对于提高空间组织和空间解读能力大有助益，尤其对于城市设计研究，这种快速图纸可以很清楚地表明城市与绿地系统之间相互嵌套的关系，以及在同样的绿地构成情况下，不同密度的城市、不同天际线的组合可以对城市意象产生的影响。

City imagery of Boston Harbor

The following drawings present two scenarios, the Long Wharf in the waterfront area in Boston and Boston Common, both being the most peculiar components of the Emerald Necklace Park System in Boston. Practicing such rapid abstractive expression of typical landscapes is very helpful for improving the abilities to organizing and interpreting spaces. Especially for urban design study, such quick drawings can clearly demonstrate the mutual-embedment of the city and its greenbelt system, as well as the influences on city imageries imposed by different skyline combinations and by different architecture density in the city, with the same greenbelt composition.

波士顿长码头滨水区鸟瞰
Aerial view of Boston Long wharf water front area

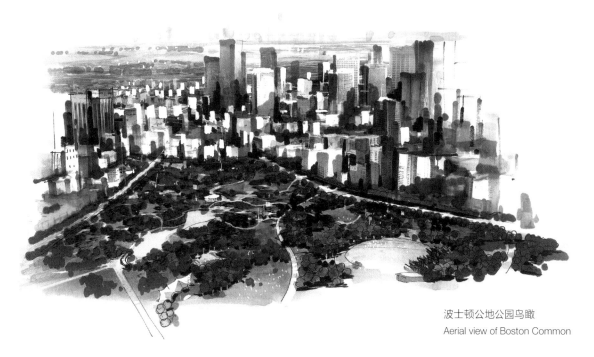

波士顿公地公园鸟瞰
Aerial view of Boston Common

083

2.6 人物的表达
2.6 Expression of People

酒吧人物宜用道具 多用人视点
Use props to depict people in the bar or casino

点景人物可用线、色块两种方式
Use lines and strokes to depict figure

表现人体运动规律
Express body movement

运动人体的基本动态要符合解剖学原理,一般人物的肩线、腰线、臀线均与脊椎线相垂直,无论动作幅度多大,只要保持三线与中线的垂直即可真实反映动作的平衡特点,不至于离真实太远。

The basic dynamics of a moving human body should comply with the principles of anatomy. That is, generally the person's shoulder line, waist line and hip line are perpendicular to the spine. Therefore, no matter how far he/she moves, as long as the three lines are perpendicular to the center line in the drawing, the balancing features of movements are truly reflected with not too much deviation.

人物配景的基本规律

　　人物配景在建筑图中主要起到尺度参照物的作用，一般不做深入的结构表现，建筑师人物以衬托环境、准确反映动态规律为限，需要了解人体运动的基本结构，如重心居中原则、三线与脊椎相垂直等原理。

　　人物黑白小稿的练习首先在动态表达，由于建筑景观配景人物目的在于尺度对比和场景性质衬托。故而，对人物细节不要求具体，但对人物动态和场景行为的表意要求较高。一般在做配景人物时需要考虑人物的动作、年龄、人物之间的配合，如父母亲带着孩子在游园、老人在看护孩子、青年人在运动以及沙滩上独特的人物与风景的配合等等。

Basic laws for expressing people as the background

People as background in architectural drawings mainly acts as a reference of dimensions, and therefore in-depth structural expression is not required. Architects shall understand the basic structure of the human body involving movements in order to use people to set off the environment and accurately reflects the laws dynamics, including the principle of centering of gravity center, three lines' perpendicularity to the spine, etc.

Expression of dynamics is the focus of practicing small black-and-white works. As people as the background of architecture landscapes is for dimension comparison and scenario property set-off, depicting specific details of them are not necessary. Instead, their dynamic status and significations of their behaviors in the scenario are more important. Often people as the background shall be expressed with appropriate actions, ages and interactions between them. For example, parents in the garden with their children, the elderly taking care of children, young people in sports, special combination of people on the beach and the landscape, etc.

人物的职业特征与场景感的表达。不同视点高度的人物表达及与光影的配合。

Expression of people's occupational characteristics and scenarios. Expression of people from trial basis of different heights and coordination with light and shadows.

所谓"世界上最美的客厅":威尼斯圣马可门前的小广场,场地的尺度和环境氛围完全靠人和鸽子的互动组成,远处教堂的入口表现极简,只是作为一个背景存在,光从海面的方向打过来,穿越总督府和钟楼之间的空隙打在建筑主面之上。人与鸽子的交错:上下翻飞觅食的鸽子、川流不息的人,突显了"最亲切的广场"这一环境主题。

The so-called "world's most beautiful living room": Venice Piazza San Marco, where the field dimensions and environment atmosphere is totally represented through the interaction between people and pigeons; the Church entrance in the distance is represented in an extremely simple way as no more than a background; light comes from the sea, pass through the spaces between the Governor's mansion and the Bell Tower to rest on primary side of the building. People interleaving with pigeons: pigeons flying up and down for food, a steady stream of people, together they highlight the theme of environment "the most cordial square".

这是一个典型的"没有设计过的广场"。如果你有幸在广场上待一天,从清晨第一个客人到来,到亚得里亚海上第一缕阳光射进来,到最后一个游客散去,而不是像普通旅游者那样匆匆一瞥而去的话,……你会发现这座建设了数个世纪、凝聚了至少四、五代设计师作品的人类空间杰作其实是"空无一物",除了共和国的几座旗杆,没有任何一点人为的所谓"景观构筑",谦逊的建筑师留下了舞台,让全世界的人都来这座舞台上,作为主角、作为人的景观去感受,这是真正的街头芭蕾。

This is a typical "undesigned square". If you are lucky enough to, rather than coming only to give a hasty glance and then going away like any ordinary tourist, spend a whole day on the square, from the time the first guest coming in the morning, to the shining of the first ray of sunshine to the Adriatic Sea, and finally to the time the last tourists going away, you will find that this masterpiece of space design, undergone construction for centuries and assembling the works of at least four or five generations of designers, is actually "nothing but emptiness". Besides several flagpoles of the country, no artificial landscape is left. The modest designers had left the stage blank, so that all human beings could come here to act as the leading role, to feel as a landscape of people. This is the real street ballet.

马克笔人物配景练习：在景观手绘表达中，人物往往是视线的焦点，对活跃场景气氛、空间尺度参照起着重要作用。画人物用笔上色通常很概括，表示出大的动态即可。采用油性马克笔可以刷出非常流畅的笔触，在小型场景构图中，前景的人物绘制对于场景真实感的塑造会产生意想不到的良好效果。

Practices of people background expressed with markers: in hand-painted expression of landscape, people often becomes the focus of eyesight, playing an important role in livening up the atmosphere and acting as a dimensional reference. People are usually painted and colored roughly, only to represent their major behaviors. Permanent markers are able to present very smooth strokes through "brushing"; in the composition of a small scene, people properly drawn on the foreground can be surprisingly helpful in creating a realistic scene.

2.7 车和船的表达
2.7 Expression of Vehicles and Vessels

　　鸟瞰的汽车往往是表现停车场和城市道路鸟瞰必不可少的道具，鸟瞰的汽车一般分前、中、后三部分，三段可依次缩进，简易型轿车只需将前后舱稍稍减小，则可表现出日系轿车的俯瞰特征；反之，在前后舱加大的俯瞰汽车中，人们一眼即可辨别出其美国大型车的某些特征。

　　Vehicles in aerial view are often inevitable in the aerial view of parking lots and city roads. They are usually divided into three sections, namely the front, middle and rear, which can be presented as being retracted in turn. For compact cars, the size of fore and rear cabins shall be slightly reduced to show the features of Japanese cars in aerial view; in the contrast, vehicles whose fore and rear cabins are larger in the aerial view can be easily recognized, because these are the distinct features of American cars.

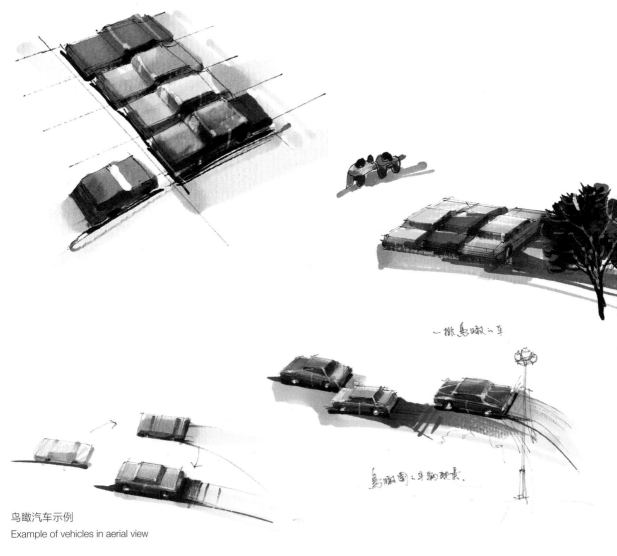

鸟瞰汽车示例
Example of vehicles in aerial view

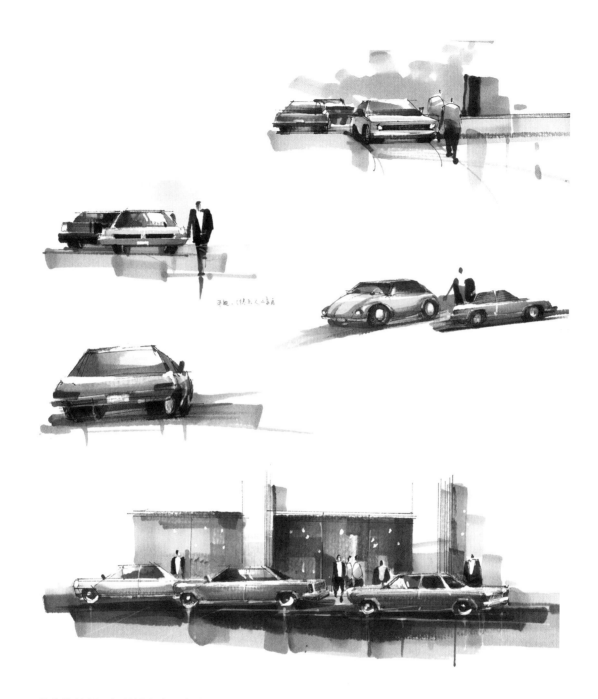

汽车的绘制一般以横向表现为主,用色上浅下深,整体比例往往夸张横向,降低高度形成低平、贴地稳定的质感。在具体绘制中,汽车的底盘往往要求稍作夸张,显出厚重沉稳之感。汽车两边的点景人物往往具有职业特征,或门童、或侍者,是突出场景感的重要元素,不可忽视。

Vehicles are often expressed horizontally, with light colors in the upper part and dark ones lower; overall proportions are always exaggerated horizontally, with the body lowered to show the sense of flatness and stability on the ground. In the specific operations, the chassis of the car are often magnified to look heavy and steady. Generally, people as landscaping elements on both sides of the car have obvious occupational characteristics, of doorman or waiter for example. They are important factors to highlight the property of the scene and can't be ignored.

在表现海港码头等场景中，往往会配以造型生动变化的船舶，色彩艳丽、造型新颖的游船或赛艇不仅能极大活跃气氛，对于特定滨水空间的表现，往往能做到事半功倍的效果。譬如，要表现水面的质感，最好是画上船身的倒影；高速进行的船只推开的涟漪，对于海滨场景的点题则最为合适。

船舶的视图表达中，视点选择尤为重要，一般需压低视线，保持视点与视平线尽量齐平，如此可以省去许多繁琐的细节，船身外侧的表现除了高高的桅杆，彩色的帆等元素，船帮两侧的旧轮胎往往是最接近于水面，另外，诸如停船的甲班，水中的木桩等，对场景气氛的渲染，也能起到独特的作用。

In scenarios of harbor, wharf and alike, vessels with varied styling can be frequently found. Gorgeously colored cruise or racing boat with unique design can not only greatly create an active atmosphere, but also double the effort to present waterfront spaces. For example, a reflection of vessel will make the water surface more realistic, and the ripples along a vessel running fast will perfectly bring out the theme in a scene of beach.

The selection of sight point is of special importance in expressing vessels. Generally a lowered sight line is required, so as to keep aligned with the eye line as much as possible. In this way many cumbersome details are skipped. Besides the masts, colorful sails and other elements high up in the air on the outer sides of the hull, the old tires at shipboards, often the closest to the water, should also be presented. Decks of stopped ships and stakes in the water can also set off the atmosphere.

3 景观色彩的表达
Chapter III Expression of Landscape Colors

3.1 植物色彩的表达
3.1 Expression of Plant Colors

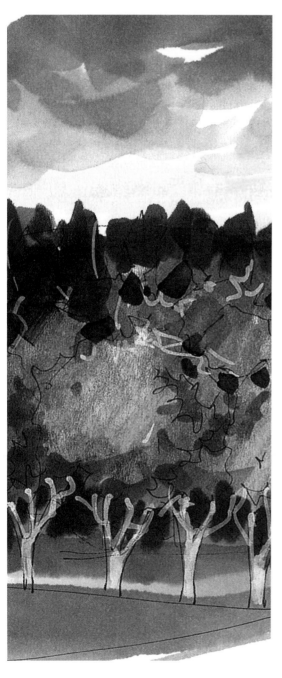

鸟瞰和透视图中的植物层次表达，通常采用深色的常绿植物作为色叶、开花乔木的背景，形成明确的前后层次区别，天空是最远、最后的层次，同样应表现得丰富而不杂乱。

When expressing plant gradations in the aerial or perspective view, dark evergreens are usually used as the background of colored leaf trees and flowering trees, so as to produce clear hierarchical difference between the front and rear. The sky is the furthest and final gradation, which should also be neat and enriched.

春、夏季节的树林：阳光下林缘的强烈对比，春日花海如烟如雾的活力、生机是表现的重点，枝干以流畅的线条"刷"出来，如雾之花、闪光之叶皆以白色涂改液直接一次涂出，形成厚度和体积感。

Woods in the spring and summer: the focus for expression shall be the sharp contrast at the edge of woods under the sun, the smoky and foggy life and vitality of spring flowers. Branches are painted in smooth lines through "brushing"; foggy flowers and shining leaves through one-off daubing with white-out, creating the sense of thickness and bulk.

某公园中轴景观廊：有如隧道一般的植物拱廊，空间深度完全由色彩对比产生。画法上仍然是彩铅底色、马克渲染退晕和白色涂改液修饰三个层次。

Landscape corridor on the axis of a park: plant arcade just like a tunnel, with spatial depth totally presented through color contrast. In respect of skill, there are three gradations as usual, respectively the color pencil bottom color, marker rendering and grading, and white-out finishing.

植物群落表现示例——松林景观

以近景的松林作为框景，构成天然的柱廊和客厅，远景的树林，线面交替，形成树林的屏障。

Example for expressing plant communities — pine forest landscape

The close-range pine woods, as the enframed scenery, constitute the natural colonnades and living room; the distant woods, with intercrossed lines and planes, form a barrier of the forest.

意象化的林间场地：用大笔触扫出深色环境背景，用坚挺有韧性的粗细线条直接刷出近景的质感，将远景的层次简化概括。

Woodland imagery: "sweep" the dark background with rough strokes, and "brush" the texture of close-range texture directly with firm, resilient lines, both thick and fine ones, and the distant gradations are presented in a simplified way.

空间表现中的植物色彩：用温和渐变的中灰、墨绿直到黑色墨线，形成符合透视效果的色彩冷暖渐变，色彩变化丰富、沉稳，天空用色纯净、轻盈，几笔刷出，与树木的浓烈色彩形成对比。左图为托斯卡纳园林中的柏树轴，下图为雅典市场的多里克神庙环境。

Plant colors in spatial presentation: lines from soft and gradually-changing medium gray, to dark green and finally to black ink, all used to produce cold-to-warm color grading with a perspective effect. The color change is rich and stable; the sky is in pure, light colors presented through a few "brushes", contrasting clearly with the intense color of the trees. The left drawing is the cypresses axis in Tuscany garden, and the below is in Doric Temple in the Athens market.

树林的表现：强调秋冬季相的树林，以枝干表现为主，强调黑、白、灰的层次关系。

Presentation of woods: seasonal changes of trees in the autumn and winter are highlighted mainly through their trunks, with hierarchical relationships of black, white and gray colors as the focus.

寥寥数笔刷出的小稿，景物经过典型化处理，笔触运用自由奔放，始于表达意象化的空间。最终形成较为抽象、意念化的景观小稿。

A small work painted through a few brushes, where the scenery is typified, with free, bold strokes used to express the space imagery, finally giving birth to this small abstract, idealized landscape work.

阳光下的农场景观：典型的景观元素如橡树、缓坡、牛群、草坪统一于长长的投影之中，投影概括了坡度和坡向。

Farm landscape under the sun: oak trees, gentle slopes, cow herd, lawn and other typical landscape elements are within the long projection, which indicates the gradient and aspect of the slope.

柏树画海.

秋色叶的林带重在纸条的疏密,线条的交角一般要顺着自然数目的长势,以锐角相交叉,密度适宜,需要着色的树林则线条宜疏,留出空间,林冠上需着意留出天空的颜色,林木间着色不宜过重,轻轻扫过,犹如空气在林中穿行,树干表达要坚实,质感与枝叶相反,粗壮的主干上宜留出林间枝条的疏影,林间道路的透视进深表达主要依靠树木投下的阴影,一般前疏后密,前实后虚,前部色彩偏暖,后部色彩偏冷。林木交错之间需要渲染出光感,这些都需要多琢磨、多练习,才能做到得心应手。

The key to express autumn-leaf woods lies in the density of branches. Generally, skew angles of lines should be consistent with the natural way of the trees' growth, forming sharp angles with suitable density. For woods that need coloring, lines should be sparse to leave out spaces. Above the forest canopy space should also be reserved intentionally for the color of sky. Coloring between trees should be moderate, done with gentle "sweeping" like air passing through the woods to avoid too heavy colors. Trunks shall be solid, a quality opposite to that of branches and leaves. On the stout trunks there should be sparse shadows of wood branches. The perspective depth of paths in woodland are mainly expressed through the shades of trees, often sparse, solid and warm-colored in the front and thick, void and cold-colored in the rear. Light should be rendered among the intercrossed trees. Exploration and practices are the only way toward proficiency.

像隧道一样的树林
Tunnel-like woods

植物群落示例——平台前的丛林

画林子，由夏到冬是一个由繁到简的变化。归结来要义只有一个——林子的氛围，体现光感，要知道阳光是无所不在的。概念性林子，画背景树时要注意透气，前景与远景中要留有天际，还要注意前景、中景与远景的过渡。

Example of plant communities — jungle in front of the platform

The painting of woods, from summer to winter, changes from complicated to simple. To summarize, only key is the atmosphere in the woods, or in other words, to show the sensation of light. You know, sunlight goes almost everywhere. For conceptive woods, leave spaces for "air permeability" when drawing the background trees. Horizon should present in the foreground and background; the transition between foreground, middle ground and background should also be taken into account.

以下数例为油性马克笔描绘的自然山林，力求用笔简略，大笔触横扫画面，画面一直保持远、中、近景层次和黑、白、灰的色彩区分，尤其是远景的天空，浓淡数笔一次刷出来，天空的色彩从明黄到纯灰色皆可，关键在于笔触本身的变化和灵动，天空云流是景观效果图中占面积最大的部分，既要有变化又忌琐碎，用色须统一，笔触宜多变，宜多练习。

Following are examples of natural mountain forests depicted by markers, where strokes should be as brief as possible. The general picture is drawn through rough brushing, with the ever-clear gradations of the foreground, middle ground and background, and distinguished colors of black, white and grey. Especially the distant sky, painted in varying shades at one time, can be any color from bright yellow to pure grey. The crucial point is the varied and flexible strokes. Sky and cloud currents account for the largest area in the landscape drawing; they should be changeable yet without many details, with consistent coloring and variable strokes that requires practices.

（1）几乎是概念性的线条结合形成的粗略轮廓。

（2）主体景观作色的同时，进一步完善造型，这一步用色需浑厚，形成强有力的前景。

（3）细节完善，画出前景大树，用白色涂改液勾勒层次，浅灰色系马克笔涂抹阴影和天空远景。

(1) Almost conceptual lines combined to form a rough outline.

(2) At the same time of coloring the main landscape, further complete the shape. In this step heavy colors shall be used to create a strong foreground.

(3) Complete the details by painting big tree in the foreground, outlining the gradations with white-out, painting the shades and distant sky with light-grey markers.

以下两幅小稿为风景区冬季景观的意向渲染图，重在整体氛围渲染，大量使用灰色系油性马克笔触，扫出白雪山坡上的光影和山峦机理，前景大树以水性笔作深入刻画，重在点睛。墨线底稿在景观的描绘中起到了骨架的作用。

The following two small works are imagery renderings of winter landscape in scenery spots, with rendering of the overall atmosphere as the focus. Gray-color permanent markers are largely used to paint the light and shadows on the snow-covered slope as well as texture of the mountain range; the foreground big trees are depicted carefully with pens as the finishing touch. The ink line draft is the framework for landscape depiction.

树的地域性需画出树的冷暖。寒带最常见的高山冷山林，是表现雪场、雪山度假区规划时最常见的场景。此类场景的自然要素，如天空、山色、林带色彩均比较灰暗。规划的雪场在冬季表现时实际只有白灰两色。松林则多表现为黑色。故天空色彩反而会显出绚烂。云天、光影用笔要活、要奔放。林下色彩反而既柔又灰。右下角多层次的山林景观，在规则表现中很常见，层次由暖到冷，远景的亮色使整个画面的色彩得到区分。对此，与通常远景用冷色退后的画法相反。

The tone of trees, warm or cold, shall be presented to show the regional features. Alpine fir woods, the commonest seen in cold zones, is frequently used to present the planning of ski resorts and holiday resorts. Sky, mountain, woods and all other natural elements in such scenes are in relatively gloomy colors. Ski fields planned with large-size are actually in white and gray only in the winter, and pine woods are generally shown as black. In such a contrast, the sky appears bright and beautiful. Clouds, sky, light and shadows shall be painted in flexible and bold strokes, while the lower part of woods is in soft and gray colors. The gradated mountain forest landscape at the right corner is common in regular presentation. Gradations change from warm to cold in colors, and the bright background divides the colors of the whole picture apart. This is exactly contrary to the common skill for presenting background, which is to push it backward with cold colors.

3.2 天和水色彩的表达
3.2 Expression of Sky and Water Colors

　　水面用笔宜活，色彩以天空颜色为参照，避免简单处理。水色表现最为简易的方式是先彩铅混涂，把水光中所需的各种色彩以同类色的原则配比上底色，再与同色系马克笔混合，水陆交界处稍加影响，最后以白色涂改液提高光。这样做既快捷，色彩也更丰富。

　　Water should be in flexible strokes, with colors consistent with that of the sky and too simple processing avoided. The easiest way to represent water is to first apply color pencils in a mixed way, match all colors needed in the water light onto the bottom color in the similar color principle, and then mix with markers of the same color family, with slight finishes at the sea-land bordering, and present highlights with white-out as the last step. This is a quick method which also makes the picture more colorful.

水的色彩表达

事实上用色彩饱和的油性马克笔同样可以表达出如水彩一般精细的水体色彩，上左图依靠色彩冷暖的变化退晕表现水波的精细变化与天空的反射，右图则用均匀的明度变化表现水波的退晕效果。天空与海水之间总是有着明显的色彩一致性和微妙的色差，岩石的表现以快捷、简约为主，主要起陪衬作用。

Expression of Water Colors

In fact, the color-saturated permanent markers can also represent the colors of water as elaborate as watercolor. In the upper-left drawing, the subtle water waves changes and reflection of sky is presented through the grading between warm and cold colors; in the right drawing, uniform variations of brightness shows the grading of ripples. Between the sky and the sea, colors are obviously consistent yet with subtle differences. As a background, rocks are often presented in a fast and simple manner.

　　画面选取了悉尼最具典型意义的景观——贝尼朗岬角作为表现核心，此处三面环海，具有极开阔的视域，著名的歌剧院和悉尼植物园均坐落于岬角之上，整座半岛犹如一艘起航的巨轮，伍重设计的风帆形薄壳建筑更加强了这一起航的意象；画面有意略去前景的另一座半岛，并将近景的游艇码头等细节以最简化的笔触加以表达，以突出港湾一线楼群和伸入海上的歌剧院。同样，两者之间的植物园也表现得极为简略，形成具有音乐感的表现节奏。画面前景的大片海水色彩丰富，可以表现天空的倒影，而疾驶而过的快艇激起的水波则进一步打破了海面的平静，并给画面带来运动感。

Bennelong Point, the most typical landscape of Sydney, is selected as the core of presentation. It was surrounded by the sea in three directions with very wide visual field. Both the famous Opera House and Sydney Botanical Garden are located on the Point. The peninsula looks just like a setting-out giant wheel, which imagery is enhanced by the sail-shaped, 5-layered thin shell building designed by Jorn Utzon; another peninsula in the foreground is intentionally omitted, and details like the marina at close-range are expressed with the simplest strokes, so as to highlight the building complex along the harbor as well as the Opera House stretching into the sea. The Botanical Garden between is also in brief strokes, another part of the musical presentation. On the foreground is a large area of colorful sea water, reflecting the sky above; the speedboat racing along brings dynamics to the calm sea.

曹妃甸曹妃湖滨水区景观鸟瞰
Aerial view of Caofei Lake waterfront area

建筑表现示例，延庆环城水系

Architecture rendering example for surrounding city water system of Yanqing county, Beijing

上海金山生活水岸多层滨水景观带设计鸟瞰
Aerial view of the multi-layer landscape area of residential water-front planning of Jinshan District, Shanghai

秦皇岛金梦海湾滨水景观带总体鸟瞰

Panoramic aerial view of the waterfront landscape area in Golden Dream Bay, Qinhuangdao

　　大海的表达要根据周边环境的需要，一般大鸟瞰的水面应与天空保持一致，稍加些岸线投影，点缀些许海浪、海鸟、船只、人群即可。这里需要特别留意的是海与沙滩交界处即浅海区的处理，黄色的沙滩与蓝色的海水相交在天光下应透出些许的绿色，这样的效果需要通过快速的运笔来产生。此外，在海天相交处不要忽视对光的表现，用黄色马克与蓝紫色系马克相配合，可渲染出逆光的景象。

　　The expression of the vast sea shall suit the needs of the surrounding environment. Generally speaking, the water surface in the aerial view should be consistent with the sky, with proper projection of bank line, and dotted with some waves, sea birds, ships and people. Special attentions should be paid to the processing of the junction of the sea and beach, or in other words, the shallow sea. The yellow beach meets the blue water in the daylight, creating something green, which should be presented through quick strokes. Moreover, don't neglect to represent the light at the intersection of the sea and sky. The backlighted scene can be rendered using yellow markers in combination with blue and purple markers.

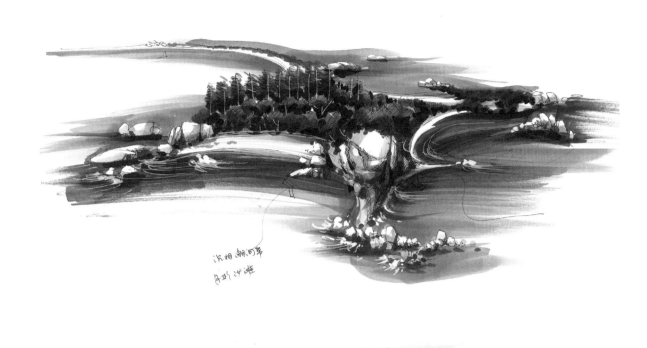

大面积水体的表达
Expression of large-scale waterbody

海水和沙滩的表现，主要注意海水的色彩过渡，通常的感受是灰到纯粹的蓝，到最深部分的前部沿海再到潮间带，混合了沙滩的颜色，简单而言就是蓝灰、到紫、到黄绿四个颜色层次的变化，在此基础上根据天空的颜色和光线的变化再有所调整。最终天空的最远部分和海的最远部分以蓝灰色相混合。前景部分是海水的蓝和沙滩的黄相混合，形成最外层的色彩过渡。

水也有节奏的变化，近海湾海水波浪的大小取决于海湾的尺度，所以在我们用白色提亮海浪的时候，浪花的前一层和后一层的距离要保持适度，浪头的密度不能前后均等，总之要体现出变化。

Pay attention to the color grading of seawater in the presentation of the seawater and beach. The usual feeling will be from gray to pure blue, to the coastal zone in the front with the deepest color and then to the intertidal zone, with the beach's color mixed in. To be simple, the color changes involves four gradations, from blue gray to purple, and finally to yellow green. On this basis, adjustments can be made according to sky's color and light changes. Finally the sea joins the sky in their farthest part in blue-gray. On the foreground, blue water mixes with the yellow beach, resulting in the color transition at the outermost gradation.

Water rhythms are also in changes. The dimension of the sea gulf determines the near the size of sea waves near it. So when we are brightening the waves with white color, keep appropriate spaces between the neighboring layers of sprays, and the density of wave trends shall also be varied to reflect the changes.

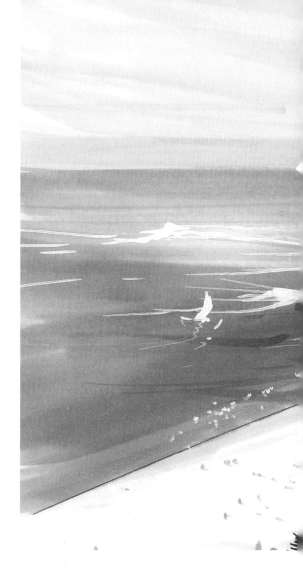

南湖中央岛配植设计，在分区景观图中特地突出了植物的季相变化：春季的开花乔木和秋季的色叶树通过不同的景观主题加以表达。中央岛上的常绿树，油松和黑松形成深色的背景，前景大量采用紫色和粉色的桃花、杏花和樱花加以衬托，色彩上采用灰色和冷色过渡，紫色为主的灰调子和纯净的蓝色水面形成对比。

In the plant design of the Central Island of the South Lake, the seasonal variations of plants are highlighted in the zoning landscape drawing on purpose: spring flowering trees and autumn colored leaf trees are expressed through different landscape themes. Among evergreens on the central island, Chinese pine and black pine form the dark background, with plenty of purple and pink peach, apricot and cherry blossoms on the foreground for setting-off; grey and cold colors are used for grading, and the purple-gray tone is in contrast with the pure blue of the water.

图书在版编目(CIP)数据

基础训练/王劲韬著. —北京:
中国建筑工业出版社,2015.12
景观设计手绘教程
ISBN 978-7-112-18979-3

Ⅰ.①景… Ⅱ.①王… Ⅲ.①景观设计-绘画技法
Ⅳ.① TU986.2

中国版本图书馆CIP数据核字(2016)第004935号

责任编辑:杜 洁 段 宁 李玲洁
书籍设计:付金红
责任校对:李欣慰 李美娜

景观设计手绘教程
基础训练
王劲韬 著

*
中国建筑工业出版社出版、发行(北京海淀三里河路9号)
各地新华书店、建筑书店经销
北京方舟正佳图文设计有限公司制版
深圳市泰和精品印刷厂印刷
*
开本:787×1092毫米 1/16 印张:8½ 字数:207千字
2017年9月第一版 2017年9月第一次印刷
定价:68.00元
ISBN 978-7-112-18979-3
(28244)
版权所有 翻印必究
如有印装质量问题,可寄本社退换
(邮政编码 100037)